Collaborative Design Survey

ELIZABETH HERRMANN + RYAN SHELLEY = RAS+E

BIS PUBLISHERS, AMSTERDAM

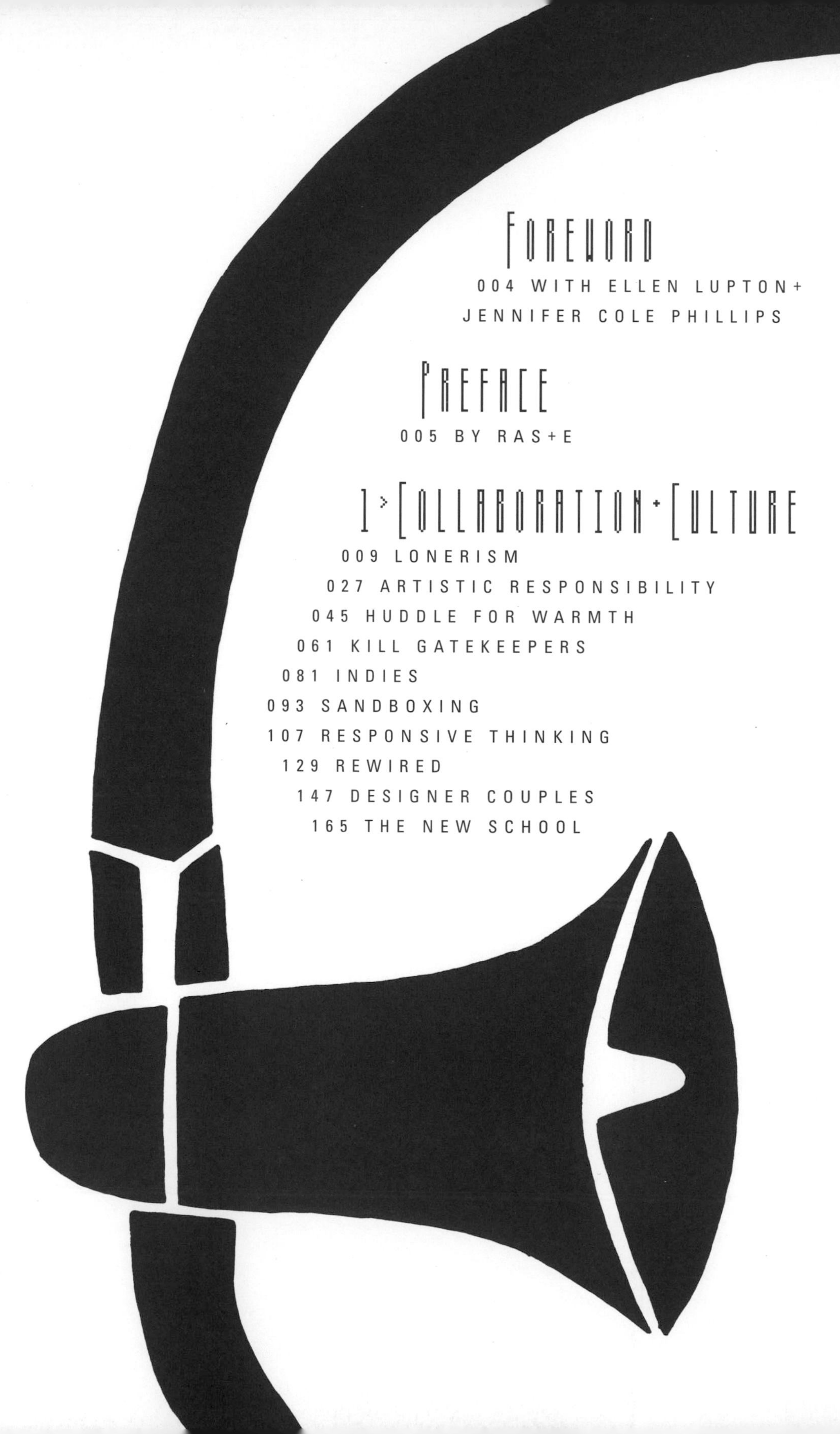

FOREWORD

PREFACE

1>COLLABORATION+CULTURE

Case Studies

Appendix

Foreword

Interview with Ellen Lupton & Jennifer Cole Phillips
Co-Directors, Maryland Institute College of Art

Do you see this Graphic Design MFA program as being unusually collaborative?

Ellen: The collaborative book project that we conduct with graduate students, like *Graphic Design Thinking* or *Type on Screen*, is by definition a collaborative enterprise. No single person would be able to do that by themselves in the time allotted, and the whole project is structured as a way to highlight the contributions of individual students.

Jennifer: In Core Studio 1, students are invited to work collaboratively, but because it's a month-long project, we feel that forcing students to work in a group for that amount of time is too restrictive.

Ellen: Too many groups are dysfunctional and we don't have time to fix them. There's always one person everybody wants to work with, the group that has no leader…

Jennifer: …There are uneven producers. Some people do all of the production while others don't pull their weight.

How do you select your candidates for the program?

Jennifer: We try to avoid drama queens—students where it's all about them. We find students that are open to a community.

Ellen: There have been collaborative thesis projects. You guys were the first. There was a collaborative the following year—

Jennifer: Kern and Burn

Ellen: —and last year, there were three guys who collaborated together. Those were great thesis projects. These collaboratives voluntarily emerged; it's amazing. I feel like you guys inspired people to see that students could do that.

What makes your co-directing work?

Ellen: MICA is unique in that Jennifer and I equally co-direct the graduate graphic design program.

Jennifer: It's not a formality—it really is a collaboration. Ellen and I bring a lot of different things to the table.

Ellen: I'm a big-picture person, and Jennifer is detail-oriented. That works well for us and the students.

Jennifer: We are both strong writers, and we love writing. That's a useful similarity because we ask that of our students. If Ellen had to do all the writing and editing, that wouldn't work.

Ellen: I am the scheduler. Being clear about who has what chore is essential to co-directing.

Jennifer: The students in our program get a lot more hands-on attention from the directors, not only because there are two of us, but because there is a strong personal commitment and investment on our part to take care of our students. I can't say any other program does that.

Ellen: I have a farmer's wife schedule. Students are covered around the clock.

PREFACE

The first ras+e project launched from an alley behind a Baltimore dive bar. An amicable drunk, displaced by last call, soliloquized our activities as we erected the prototype: a collapsible 3D cardboard missile, complete with suitcase-handle, and designed for widespread guerrilla-distribution throughout the city. Offering assistance, our chatty new friend climbed onto the tavern's roof, and flew the missile to ras, while e photographed.

The MICA GD MFA program is a risk-promoting safety net to test the inherent *maybe it will work* aspect of all collaborations. We brazenly abused democratic tools—like treating cyanotype as a printmaking medium with photocopier negatives—and discovered design solutions embedded in cross-disciplinary dialogue, an un-billable essential without parallel in the assembly-line production of ad agencies. As designerauthorfacultyimps, we see this Millennial Collaborative Ideal reflected in our peers and students.

One hundred diverse practitioner interviews became our research, an argument that small interdisciplinary collaboratives can make anything, provided old world authorship is shelved with the Lost Ark. The freestanding essays in this book are intended as contextual encouragements for young designers, forming a model for varied collaborative processes. Each Q+A hits concisely, with additional bits available online.

The visuals reference collaboration: Schizocourier is a bastard mashup of the famed monospace, buckets of red draw the line between black and white, graftedillustrationguffaws reference found works, and the grid slams itself.

Collaboration Now is Design Punk.

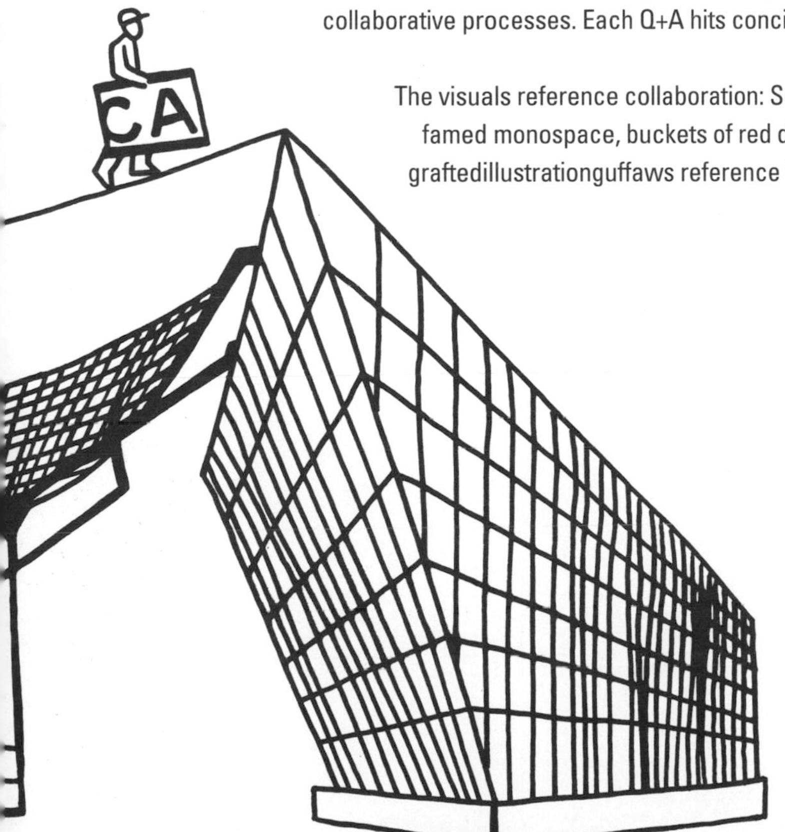

SECTION

Small design collaboratives have been gaining traction in a field historically dominated by large ad agencies and rockstar isolationist Mad Geniuses. The chapters in this section tell the How and Why of the shift: its response to cultural change, and the reciprocating impact.

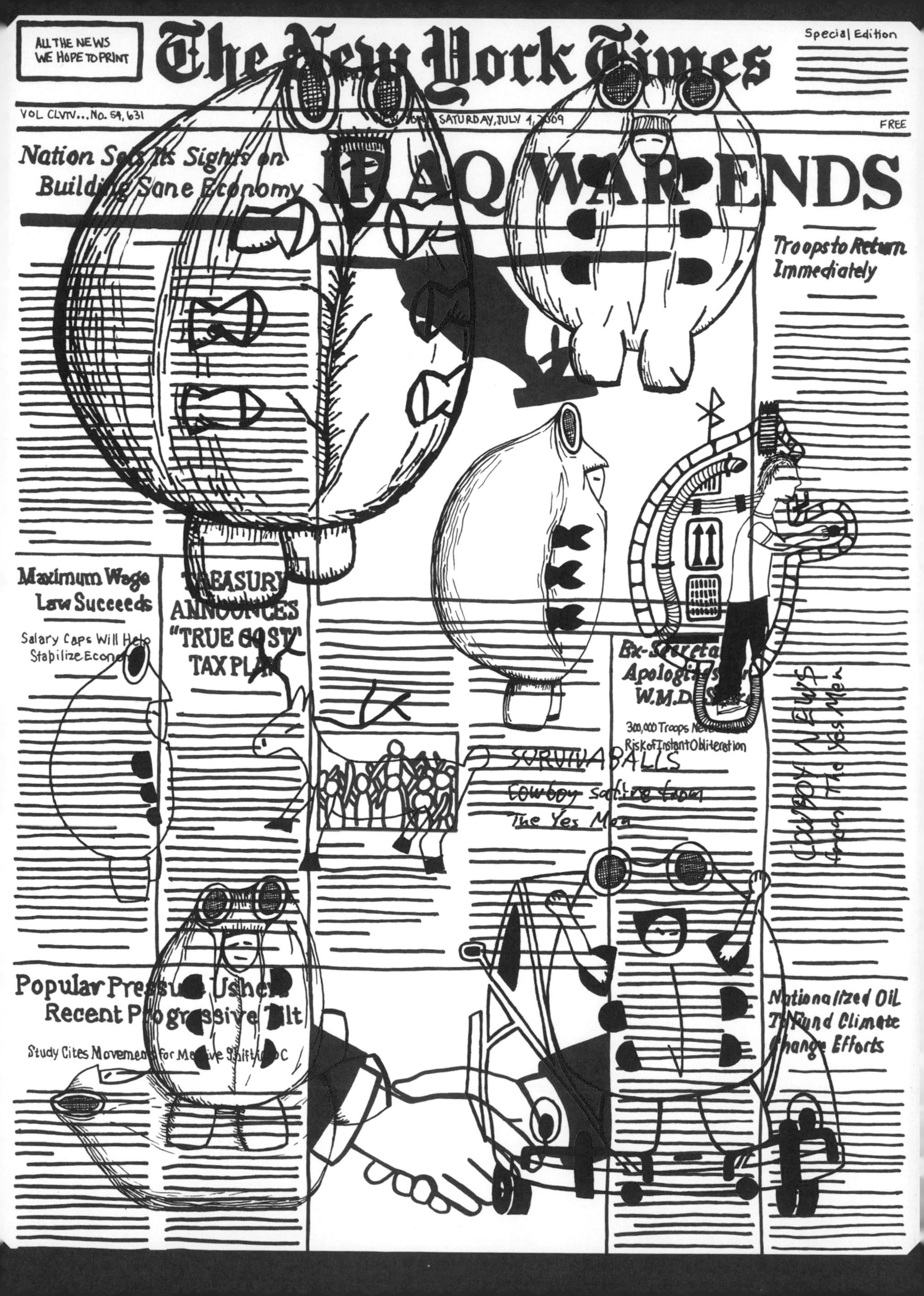

ALL THE NEWS WE HOPE TO PRINT
The New York Times
Special Edition
VOL CLVTV... No. 54,631
SATURDAY, JULY 4,
FREE
Nation Sets Its Sights on Building Sane Economy
IRAQ WAR ENDS
Troops to Return Immediately
Maximum Wage Law Succeeds
Salary Caps Will
TREASURY ANNOUNCES "TRUE COST" TAX PLAN
Apologizes
W.M.D.
300,000 Troops
Risk of Instant Obliteration
SURVIVABALLS
cowboy satire from The Yes Men
COWBOY NEWS from The Yes Men
Popular Pressure Ushers Recent Progressive Tilt
Study Cites Movements for
Nationalized Oil To Fund Climate Change Efforts

Lonerism- Death of the Mad Genius

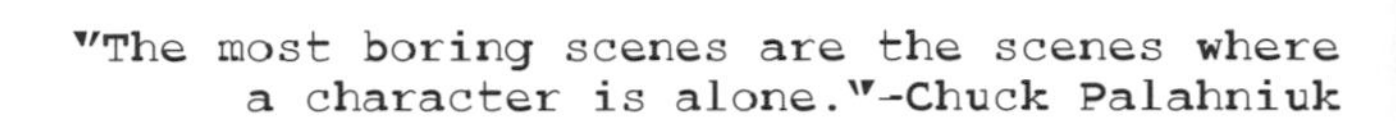

"The most boring scenes are the scenes where a character is alone."-Chuck Palahniuk

Two dozen businessmen of the climate-threatening professions waddled about the East River encased in SurvivaBalls. These inflatable life-support shelters are personal, expensive protection capsules-convenient Swiss Army body barriers against environmental disaster. The bubble suits float, crawl, recycle liquids, and naturally migrate into colonies for strength in numbers. If it weren't for the fact that you need a permit to SurvivaSwim in the East River, surviving would be totally legal.

In other news, the free July 4, 2009, Special Edition of *The New York Times* (handed out on November 12, 2008) greeted the city's denizens via volunteer town criers, who distributed 80,000 copies of near-future wishful news: George W. Bush indicting himself for war crimes, national healthcare's arrival, and the end of the war in Iraq. Working with the Anti-Advertising Agency and other volunteers, The Yes Men executed a "build it and they will come" approach to how the world could work.

Labeling themselves "identity correctors," The Yes Men take on giant corporate abuses using fake websites and cheap suits. They are the unemployed designers that parents worry about when arguing against art school enrollment. The Yes Men draw attention to human desperation and lax environmental policy while humiliating ludicrous politicians: design for the greater good on a "meta" scale.

Yes Men Andy Bichlbaum and Mike Bonanno are culture-jamming Cowboys.

Collaboration within design and the fine arts is not new, but there is a trend of small collaboratives that have a highly integrated process. Design has historically considered collaboration as assembly line production, to each his own task. The new collaboratives, a diverse and interdisciplinary group with several historical precedents, often maintain their outsider status while developing work in and for a community.

RAY JOHNSON
design cowboy

Ray home phone 516-676-3150

NOT worth your trouble to
unless you

YOU ARE INVITED TO A NEW YORK CORRESPONDENCE SCHOOL MEETING

DECEMBER 17
2-5 PM

TRUMAN GALLERY
38 E 57 ST NYC
212-688-3516

I will be directing a three hour meeting

Ray Johnson + The New York Correspondence School

Ray Johnson swan-dived off a bridge. It was a brisk swim, as most winter dips around Long Island tend to be. Only a couple local residents witnessed him casually back-stroke permanently out to sea. Police found the body the next day, on January 14, 1995, washed up on shore in Sag Harbor. They promptly entered and combed through his home studio in Locust Valley, unknowingly uncovering Ray's final elaborately planned performance piece: a carefully curated network of mindfully experienced artwork, organized as a conceptually guided tour for the intruders. The work was linked through personal narrative, launched via suicide, and cultivated within a Lonerist Mad Genius myth. King me, Chris Burden.

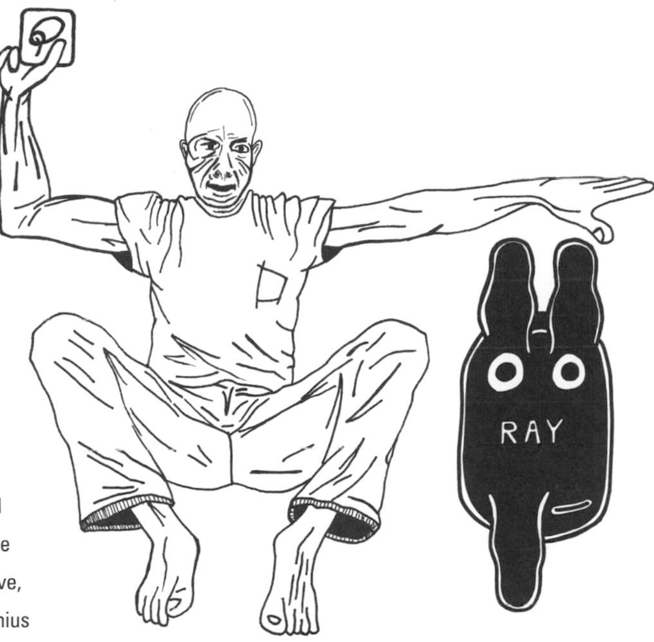

9.29.93 BILL,
PLEASE MAIL THIS
ENVELOPE WHICH
CONTAINS FOUR
●●●● ALSO THE
MOUSE TO CHARLES
FAHLEN, 945 NORTH
RANDOLPH ST, PHIL.
PENN. 19123

"New York's most famous unknown Artist," so labeled in the Ray Johnson documentary *How to Draw a Bunny,* regularly sprinted ahead of the looming trends of Pop Art, Mail Art (Interactive), Performance Art, and Intermedia. Ray was a prominent contributor to New York's art communities, with his Fluxus companions building upon his Happenings/Nothings. His repurposed collages, which he called "moticos," prophesied Pop Art, alongside two books he collaboratively self-published with Andy Warhol before Warhol became a rockstar. How Ray thought about systems, interactive art, and graphic design is exemplified by a portrait series detailed in *How to Draw a Bunny.* After drawing a line silhouette of the sitter/patron, a collection of collages were shown to the subject for potential purchase. A series of written correspondences negotiating the price quickly escalated into an overhauled "mathematical" system of additions and subtractions. Ray often priced work collaboratively. Photocopies of the evolved works were mailed to the patron in response to various offers and changes. Once the patron received all the photocopies, Ray's portrait was finally complete. The final output treated the initial collages as a mere launchpad. Ray's Correspondence Art completely subverted expectations and historical conceptions of portraiture. Clearly more interested in the process of constructing an elaborate system for rhetorical communication, the negotiations created a photocopied portraiture set that became the true series.

So, is Ray a Loner?

To revisit our earlier "Lonerist" allegation deeply and accurately, we must unpack and thoroughly excavate the wrought layers of the "Ray Johnson" collage. Ray's relevance to the collaborative design discussion extends beyond his work as a graphic designer and systems thinker. Whether it was the influence from his student days under Josef Albers at Black Mountain, or working in New York, he made various art communities the subject and focus of his communications. Ray was a leading revolutionary of western Cowboy Artists. Ray was an Outsider, but not a Loner. He was the prototypical Artist as Cowboy Hero who cared deeply about the community, though he was not of it. Ergo, Ray possessed perspective and created with altruism. His later work from Long Island paralleled his withdrawal from the public New York scene, while fostering an enlargement of the Mad Genius myth as a mechanism to subvert. Furthermore, Ray's work overtly addressed the dichotomy of detached physical correspondence, foreshadowing a looming and all-encompassing technological presence. Leaving his friends in NYC and corresponding from Long Island was a tee for his blossoming interactive work, reinforced by a large increase of mail contacts. Although suicide can read like a Lonerist act of an isolated Mad Genius, Ray saw it as the opposite: a specific, controlled farewell dedicated to an increasingly disconnected telecommunity.

Mail bombs away!

Outsider Cowboy Heroes vs. Mad Genius Loners

Cowboy Hero Trope slid off his horse, driving his boot-heels into the swirling street dust, and immediately marched forward. Confidently, he banged aside the saloon doors, dark eyes darting as his right hand softly twitched over the heel of his gun grip. Sixty pairs of light-adjusted eyes tracked his intrusion, noting his inelegant leathers and sun-battered cheeks. The pistoled outsider shrugged off the townie gazes and leaned across the counter, signaling for the bartender's attention. An unattractive gravel voice queried the drink slinger: "D'ya know who asked for a sculptor?"

SILLY
HATS
ONLY

SILLY
HATS
ONLY

ARCHITECT AS
MAD GENIUS
Liz Diller takes NYC

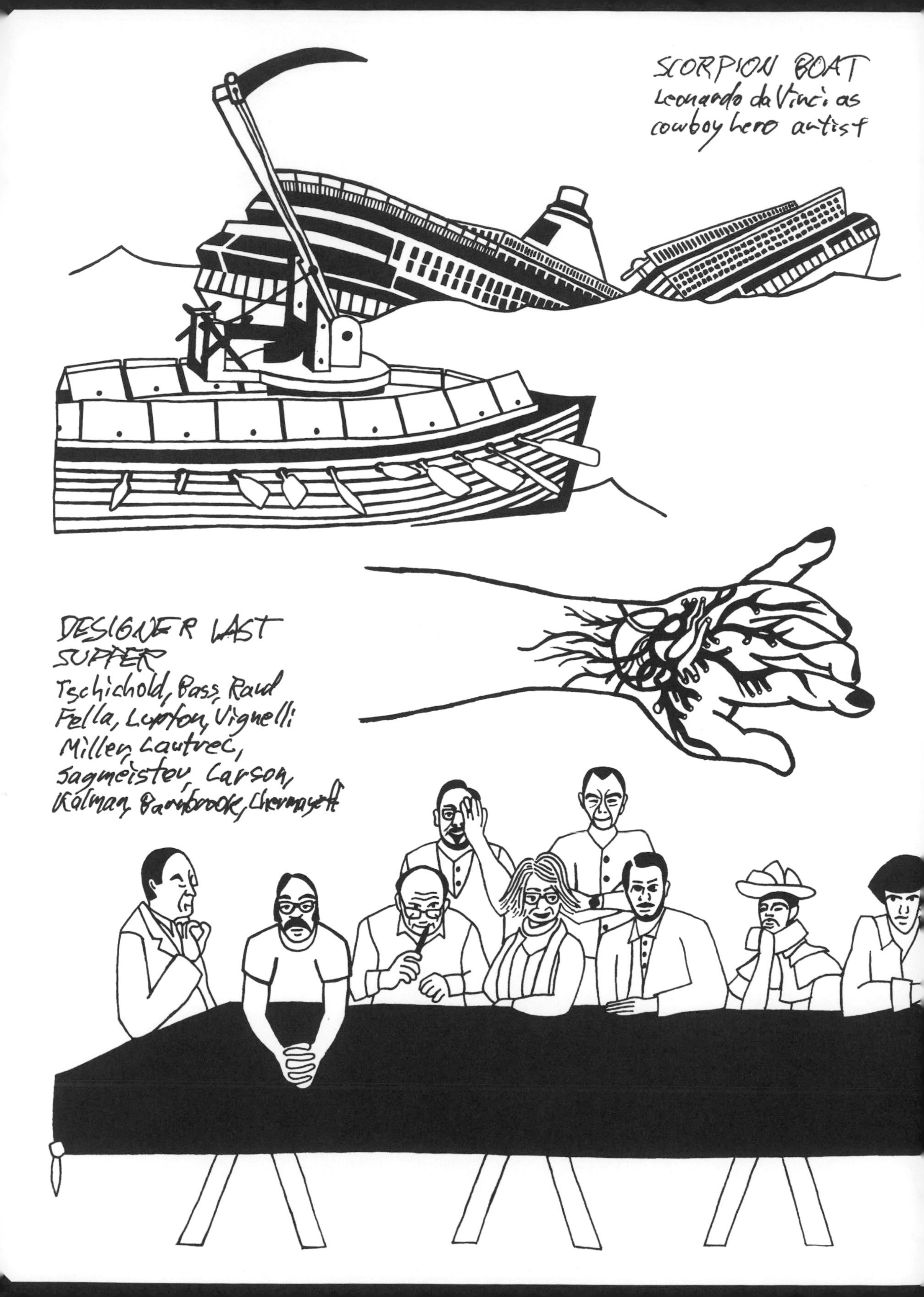
SCORPION BOAT
Leonardo da Vinci as
cowboy hero artist
DESIGNER LAST
SUPPER
Tschichold, Bass, Rand
Fella, Lupton, Vignelli
Miller, Lautrec,
Sagmeister, Carson,
Kalman, Barnbrook, Chermayeff

Enter Michelangelo. Low-class tradesman, amateur Dante enthusiast, God's sculptor and painter: the prototype for Artist as Mad Genius.

Although Michelangelo tamed the whims of nine popes on the strength of his artwork alone, he was later excommunicated from the Church, forced to burn his possessions, and flee his home. Two biographers wrote his story during his lifetime. Michelangelo's laser focus on work came at the expense of any semblance of hygiene and, in his later years, he was known to deploy a combative personality. It's incomplete to call him an acerbic Loner, but he unintentionally initiated a template for the public's perception of Artists as tortured, isolated, Mad Geniuses. Michelangelo and many other Artists of similar dedication have been retroactively diagnosed with various mental illnesses, especially Asperger's. While Michelangelo wasn't as consistently isolated or crazed as was popularly believed, many familiar perceptions of Artists originated with him. Says Michelangelo biographer William Wallace, "He did more than any other Artist to raise the stature of the artistic profession. It was a craft before him, and it becomes a profession of geniuses after him. Now, Artists have a certain dignity in society even if they are still considered marginal weirdos."

Enter Da Vinci. A contemporary frenemy of Michelangelo, he established a comet parallel to Michelangelo's brilliance. Da Vinci as an Artist is associated primarily with two paintings, *Mona Lisa* and *The Last Supper*, but most notably for his massive trove of inventions and sketches. While Da Vinci was the prototypical Renaissance Man, it was Michelangelo as Artist who was labeled *Il Divino* (The Divine One).

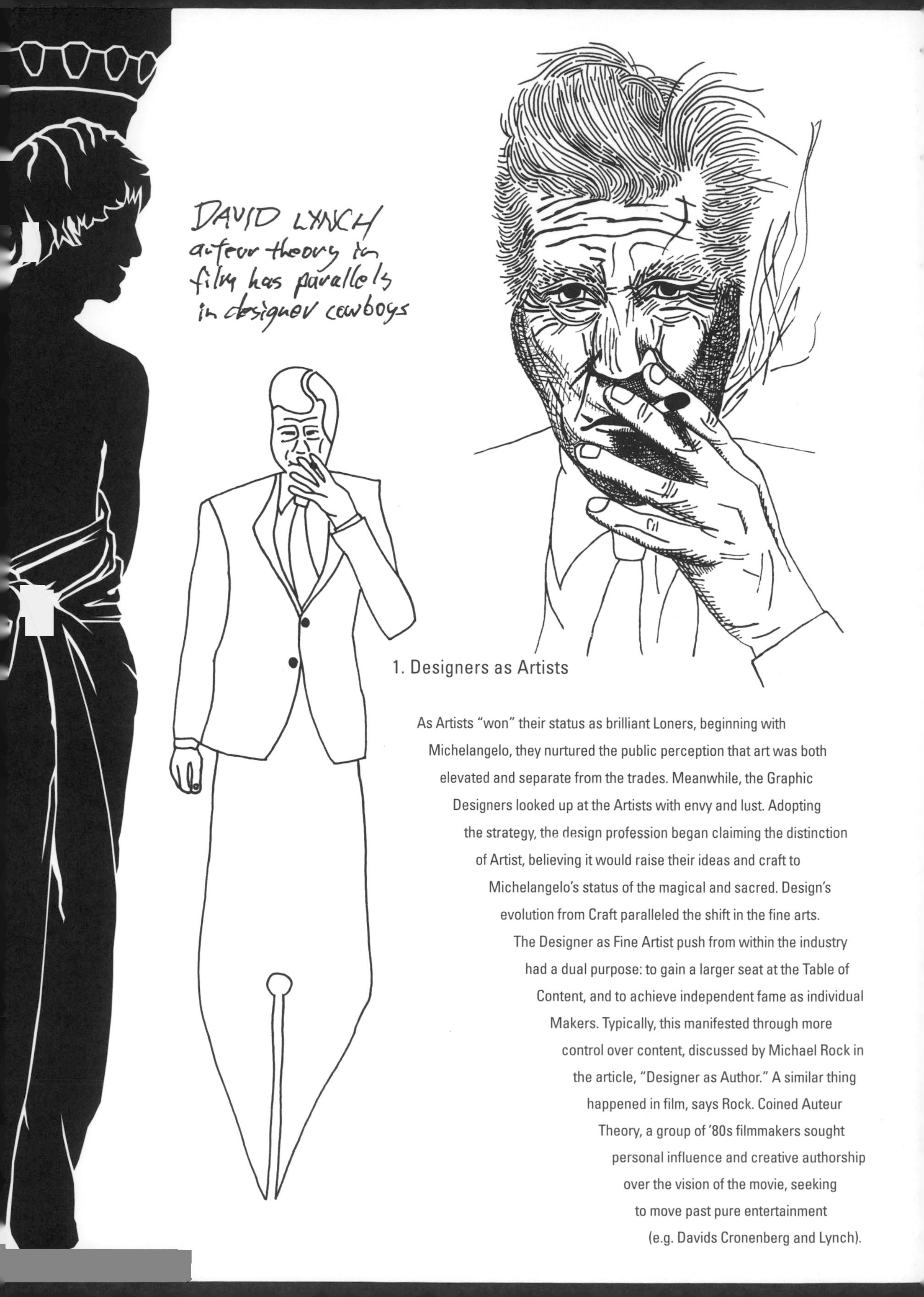

1. Designers as Artists

As Artists "won" their status as brilliant Loners, beginning with Michelangelo, they nurtured the public perception that art was both elevated and separate from the trades. Meanwhile, the Graphic Designers looked up at the Artists with envy and lust. Adopting the strategy, the design profession began claiming the distinction of Artist, believing it would raise their ideas and craft to Michelangelo's status of the magical and sacred. Design's evolution from Craft paralleled the shift in the fine arts. The Designer as Fine Artist push from within the industry had a dual purpose: to gain a larger seat at the Table of Content, and to achieve independent fame as individual Makers. Typically, this manifested through more control over content, discussed by Michael Rock in the article, "Designer as Author." A similar thing happened in film, says Rock. Coined Auteur Theory, a group of '80s filmmakers sought personal influence and creative authorship over the vision of the movie, seeking to move past pure entertainment (e.g. Davids Cronenberg and Lynch).

AIGA MEDAL

The more the myth grew, the more involvement designers had in all facets of campaigns, production, and promotion. Extending beyond Arbiter of Taste, Designer as Author involved both design and content generation, but most importantly, it implied the divine. Paul Rand was/still is a rockstar: his moves, the name, and Hero-worship. His work is sex on the beach, and his contribution to design history is impervious to time. David Carson also grabbed the mantle and took to the stage, oozing attitude and proselytizing, while New Wave followers celebrated his backwards-reading neon matinée signage. These heavy-handed quintessential figures attracted clients and the Mad Genius myth grew.

Designers as Artists are assumed to *know* things and have contagious, profitable personalities tied to their stardom. They claim seats at the table because they advertise a distinct, inherent voice (ignoring Beatrice Warde's *Crystal Goblet* advice), transcend trendy design influences, and answer directly to angels: *Il Designo*. But theatrical stages have an isolating factor, making these practitioners immune to criticism, but also to collaboration.

RAS+E CARYATIDS
designer as author
author as architect

2. Designers as Not Artists

Later, it became fashionable for Mad Genius Designers to reject the Designer as Artist model, eventually considering the two fields equal but different, which Michael Rock recognized and clarified in a follow-up article, "Fuck Content." Designers should not need to be anything but Designers. Separate but equal. This perspective somewhat cracked the Designer as Mad Genius perspective, leaving design without the clout of Architects or the exploratory interdisciplinary nature of contemporary Fine Artists. Design no longer needed anyone's approval, or influence.

3. Designers as Collaborative Cowboys

In the meantime, Fine Artists were outright ditching the isolationist Mad Genius myth, climbing off Lonerism Tower and engaging each other as remixers, commenters, and generative Interactive Artists. The collaborative nature of contemporary Community Art, the heroic democracy of Street Art, and a mounting acceptance of practicing collectives as Makers themselves were enabled by communications technologies and supported by a global culture shift.

Before telecommunications media, Fine Artists usually came together for two main reasons: (1) geographical proximity, and (2) like-minded stylistic pursuit. The reason behind these gatherings wasn't necessarily for collaboration (such as Cubists Picasso and Braque), but for purposes of Show and Tell or apprenticeship. Consider Realism, a movement originating in France directly following the 1848 Revolution as an open rejection of Romanticism's open emotion. Courbet, Daumier, Millet, and Manet aimed to present ordinary life as a transcribed reality based on truth. Branches of Realism formed in Russia, America, and elsewhere; however, these chapters formed under very different circumstances, and the context of the paintings directly related to specific environments. Many movements in art began as loose collectives, attractive simply as a platform for reciprocating critique and dialogue. Salons and night cafés supported rising local Artist Collectives within a community through physical working and meeting spaces. Many guests interacted as well, enabled by the public format of the gathering.

Starting in the '50s, telecommunications media undermined the requisite of geographical proximity in qualifying Avant Garde art movements and gatherings.

WANTED

DEAD OR ALIVE

BANKSY

$5,000 REWARD

NOTIFY– Marshall Pat Garrett

BANKSY AS COWBOY
London has an uneasy relationship with its illicit outsider-hero

KEITH HARING
activist street designer
Philadelphia, PA

"THE THING I HATE THE MOST ABOUT ADVERTISING IS THAT IT ATTRACTS ALL THE BRIGHT, CREATIVE AND AMBITIOUS YOUNG PEOPLE, LEAVING US MAINLY WITH THE SLOW AND SELF-OBSESSED TO BECOME OUR ARTISTS. MODERN ART IS A DISASTER AREA. NEVER IN THE FIELD OF HUMAN HISTORY HAS SO MUCH BEEN USED BY SO MANY TO SAY SO LITTLE."-BANKSY

MIRKO ILIĆ
designer cowboy

SAFETY FIRST
censoring cowboy journalist Spider Jerusalem from Warren Ellis and Darick Robertson's Transmetropolitan

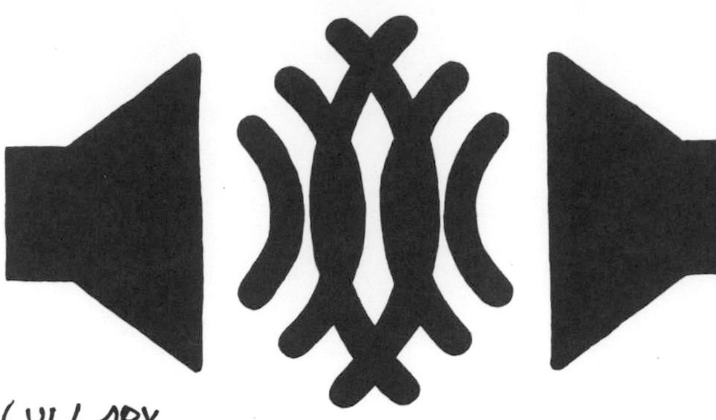

LULLABY
Chuck Palahniuk takes on noise pollution

Along with Ray Johnson, artists like Nam June Paik and, more recently, Brian Wood *(Channel Zero)*, posed commentary on the effects of television on American culture. Isolationist mythology began to erode. Chuck Palahniuk wrote in *Lullaby* that "people used what they called a telephone because they hated being close together and they were too scared of being alone." Communications technologies enabled a more broadly based collaborative capability, though working in proximity declined.

As usual, design mirrored art's shift. While some rockstar Designosaurs perpetuated the Mad Genius Lonerist activism through posters donated from on high, many contemporary teams responded as small, informal, interdisciplinary, and socially-aware groups on the ground, possessing more encompassing altruistic agendas. These design collectives rejected mythology and discipline boundaries while retaining outsider status, hence the Design Collaborative as Cowboy. Not that *all* outsiders are altruistic cowboys, but they often are. Designers exist in many roles, but "hired guns" covers many of them. They can take aesthetic risks, make hard ethical decisions, and collaborate with an aggressive voice—all without elitism—through the privileged lens afforded by being outside of the client's world.

"People used what they called a 'telephone' because they hated being close together and they were too scared of being alone."
Chuck Palahniuk

Re-enter Outsider Cowboy Hero Trope. R. Phillip Loy's *Westerns and American Culture, 1930–1955* describes one of the primary ways in which the Cowboy Hero relates to a community: "He was not part of the town, but an outsider who rode away when his job was finished." Cowboys are called upon when the town gets itself in trouble, usually the result of no exterior perspective. Cowboys are day-saver outlaws. Loy stresses that while a Cowboy Hero is an outsider with unique perspectives and skill-sets due to the nature of his job, he is not a Loner. Maybe lady folk don't like their pies getting shot to death on the windowsill during target practice, but Cowboys are appreciated, respected, and make friends despite any messy outsiderness. Cowboys cannot belong to a group; if they did, they would no longer be outsiders, their eyes seeing only the same things all other eyes see. What makes a Cowboy useful comes out of an altruistic set of beliefs, lifestyle, habits, and abilities that don't fit into the community. For example, *Adbusters'* Kalle Lasn and illustrator Mirko Ilic place their work and ideas into service, refusing the Lonerist crown of the Mad Genius while nurturing an outsider perspective.

NAM JUNE PAIK
Electronic Superhighway
tuned to BRIAN WOOD's
Channel Zero

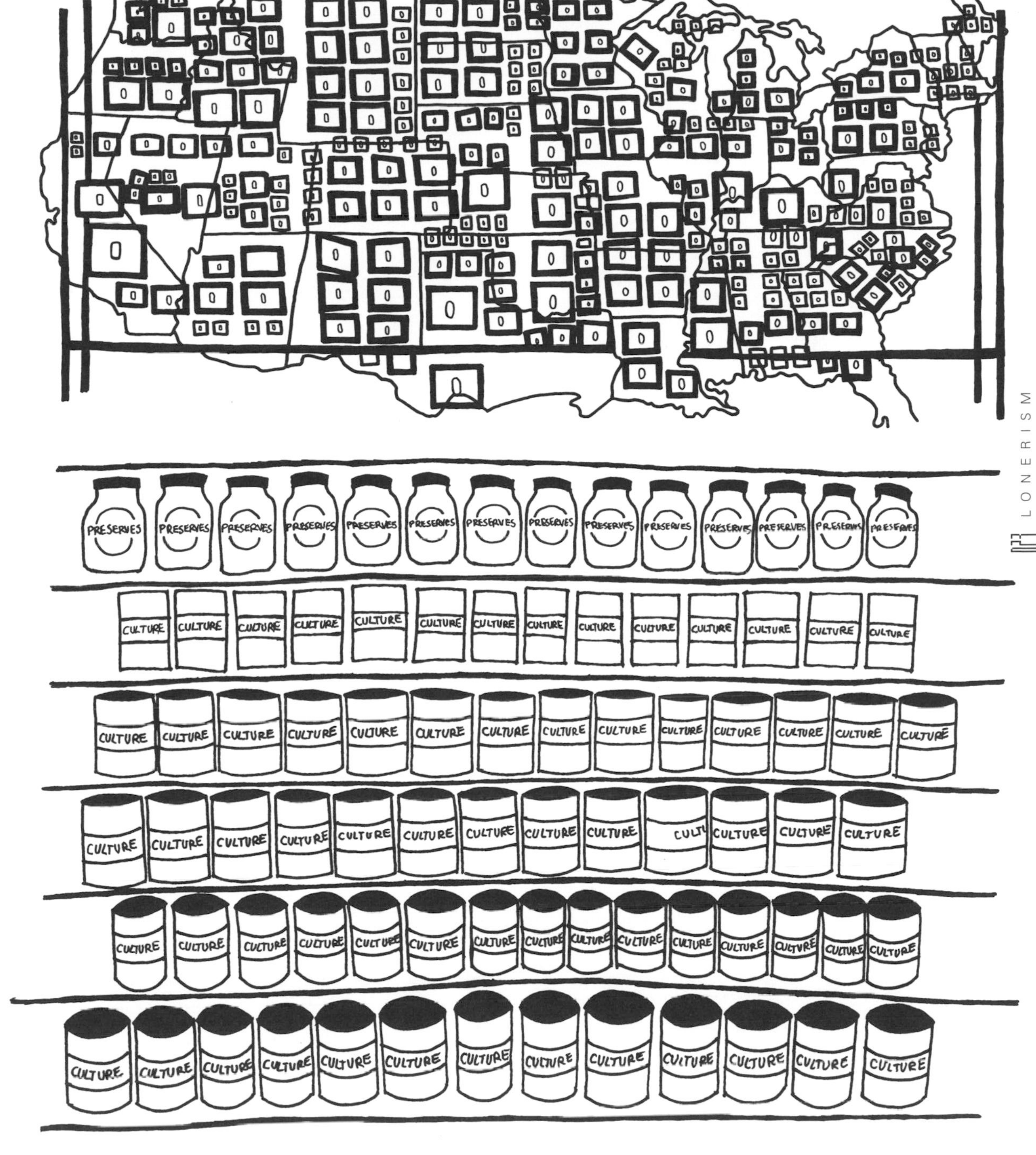

CULTURE CANNING
remixers' fear begat
culture jamming

DEXTER AS COWBOY
remixed from poster
by Shepard Fairey,
another cowboy
POWER-SAW TO
THE PEOPLE

All Together Now

Historically, designers sought respect and clout from clients by separating themselves from other practitioners and tradesmen. The argument sounds something like: design has always been collaborative; or, designers have to at least tolerate interaction with proofreaders, art directors, editors, photographers, and, sometimes, peers. But this is merely assembly line interaction. A inserted into B, then B inserted into C, and the result is a model plane. Or, a poster advertisement for a model plane.

When design is production-oriented, solitude can aid focus and efficiency. The newer collaborative model embraces process, overlapping skills, and experimentation over production. It is also less managerial and a more discursive relationship than in large ad agencies. Combined results of this design democracy foster a need for concept-driven, idea-first joint authorship to rise above the allure of Lonerist Geniuses. The idea of a Cowboy Collaborative is to form a gestalt, where the sum is greater and more unique than the individual parts: A x B = RED. The result arises from the unexpected collision of its component designers.

THOM YORKE
Radiohead's cowboy,
frontman remixed
with SANDIN YOUREYE
project The Fallen

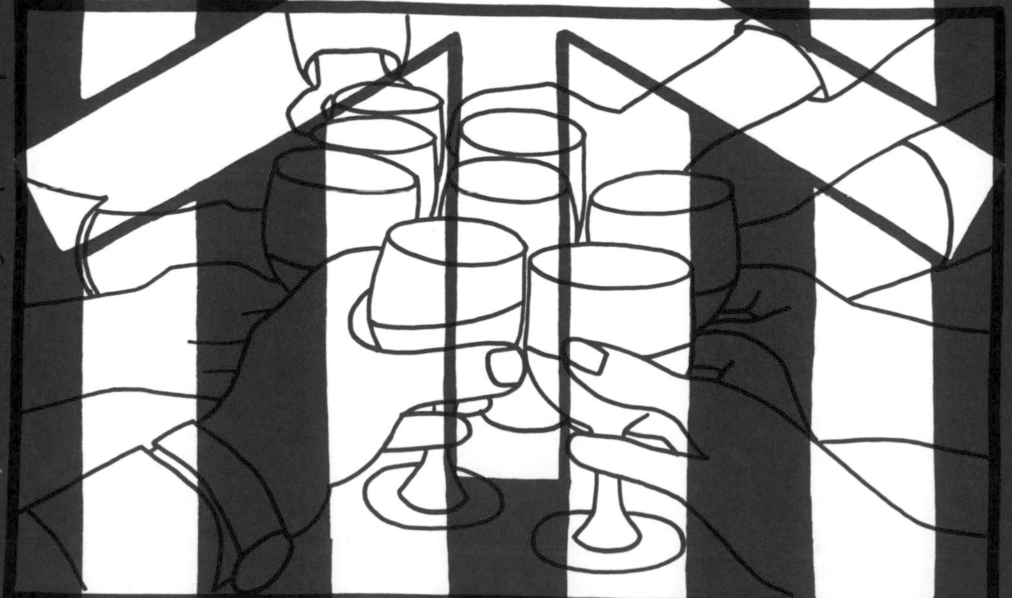
MARIO AND THE MAGICIAN
Thomas Mann's critique of Italian Nationalism and advocation of artistic responsibility
DEMOTIVATIONAL
rewriting Despair, Inc posters
COLLABORATION
NEVER UNDERESTIMATE THE POWER OF STUPID PEOPLE IN LARGE GROUPS

Artistic Responsibility

"What good would politics be, if it didn't give everyone the opportunity to make moral compromises."
-*The Magic Mountain* by Thomas Mann

An average German couple is on a vacation in an average Italian coastal village. The parents let their 8-year-old bathe naked while they wash the child's swimsuit, and the nearby locals are not impressed. Airing out their conservative nationalism all the way to the police station, the townsfolk press the issue until the family is fined and sent back to their hotel full of disquiet.

Cipolla, Thomas Mann's titular sorcerer in the 1929 novella *Mario and the Magician*, is a Ghost Rider-level antihero, a badass bad person, a cripple with superpowers. He descends on the average Italian coastal village with a whip and small tricks. The magician stages a show and the whole town shows up to see the touring "artist," including the visiting German family. After a slow beginning, the magician settles into a noticeable groove, an excellently crafted performance, his starts, stops, and his reading of the audience is impeccable. The ante pulses and grows, matching the magician's increasingly aggressive and sardonic tricks and hypnotisms. Cipolla gimps and snarls and whip-cracks the viewers in time to his debilitating internal power struggle, while silver-tonguing the sore spots left by his controlled demeaning of the audience participants. A true master, the artist organically and subtly raises the stakes of each manipulation until the meek waiter Mario is called to the stage. The sorcerer flays open Mario's deepest longings by assuming the identity of a local girl that Mario secretly adores, and under hypnosis, Cipolla coerces Mario into a kiss.

The magician's carefully built emotional wave, the slowly woven hypnosis of the room, all shatters in sudden gunfire. Wakening from his trance and realizing his absolute humiliation, Mario instinctively reacts, draws a gun, and Cipolla drops dead on stage.

BITCHES
AINT
SHIT

Ben Folds remixes Dr. Dre, vas+e remixes Jessica Hische

Patton Oswalt, Lester Beall, Richard Serra, Warren Ellis, Luba Lukova, JR

Competition vs. Collaboration: My Cause Is Prettier Than Your Cause

Art school frequently references Artistic Responsibility. Design school, semantics aside, not as much.

When designers do talk about *responsibility,* it tends to do with materials (going digital or those brown envelopes), re-illustrating Ronald McDonald with a flaming beard in between real projects, but for free (Jonathan Barnbrook's *Rosama McLaden),* or through design's equivalent of the local Episcopal youth group on a two-week trip to fix a city someplace adventurously exotic (poster donation for relief efforts). In short, charity. These historically relevant themes of design responsibility have much to do with links to significant cultural trends, and they acknowledge the warnings presented by Mann regarding art as a double-edged implement of cultural change (or Google's mantra, "Don't Be Evil"). Before developing responsibility in relation to collaboration, it is important to identify opportunity for design's increased contribution to the greater good.

McLADEN tribute to Jonathan Barnbrook

The Materials thing is ongoing. Sure, we all love French Paper. It's loud, it's eco, it's small, there's Charles Anderson; French is one of us. Nobody talks about the print industry in the same spittle-filled sentences as the automotive industry, but, not long ago, graphic design had a relationship crisis. The crisis is more ambiguous, the moral responsibilities less overt. Even in academia, legacies of *Save Paper* footnote syllabi, even in courses like *GRA4305 Yay The Type Is Moving!* Anymore, graphic designers are not first in line at the foot of the environmental apocalypse guillotine. We all want to be responsible with materials, and we are one emissions edict away from rationed materials. But there is a trap for designers, an assumption that technology will create perfect, wasteless tools. Post-print, Mass Market conflicts less with Beautiful Object from an Earth perspective—although we probably have enough first generation Kindles and iPads to shingle leaky roofs in third-world countries. And yet, not being evil does not equate to contributing positively.

THE ARTIST AVENGERS

unofficial slogan

Google

don't be evil

Google Search

I'm Feeling Lucky

TOMS

BOBS

BOBS

CHARITY
COMPETITION

LESTER BEALL
posters for REA, pro bono

YELLOW BIRD
PROJECT
works with indie
bands to design
and print t-shirts
that are sold to
benefit charities

first

Next in Classical Design Responsibility: pro bono work. Many designers have thought of this in terms of content decisions: specializing in environmentally sound design, donating a percentage of time to important projects and causes that lack budgets, or refusing to design way-finding systems for an authoritarian government. Like practicing kinetic typography in After Effects for the Girl Effect, like donating show posters for Indie Baltimore theaters, and like North Korea. Respectively—for example.

The pro bono side of responsible design practice is trickier to comment on since each designer and each agency often exists in radically different spheres. While some agencies institutionalize pro bono work, or require designers to allocate time and talent toward need-based causes, many practitioners lack the financial or structural footing for the same commitment. One payoff for loftily idealistic design work: awards.

A Michael Bierut essay criticizing the rockstar-laden *First Things First 2000 Manifesto* argues that anything and everything should be well designed, and that creating an Industry Standard Ideal dependent on the Luxury of Time, Luxury of Client, and Luxury of Content belittles the very nature of Design: The Profession. A significant tenet of many design projects involves partnering with others to effectively communicate messages that transcend the designers' personal backgrounds. From a collaborative stance, building a link between specific elements of the design community and The Greater Good creates inequality in our Farm by bringing heaven closer to some than others, rendering team concepts moot when the members are strewn across different tiers of the victory podium. Regarding Bierut's argument and awards, the concern seems somewhat more applicable to graphic design than, say, product design, where highlighting toilet brushes is disturbingly common, vs. poster design awards that involve Charmin as a client. Rather, graphic design for museums cleans up at awards. This raises a question of whether product design, like architecture, trends toward a more content-agnostic industry compared to graphic design; but Bierut argues that everything deserves quality design and equality in acknowledgment. In other words, responsibility

LANDMINE KETCHUP PACKETS
Nick Worthington at Publicis Mojo for Campaign Against Landmines (CALM)

DESIGN WON'T SAVE THE WORLD

GO VOLUNTEER AT A SOUP KITCHEN, YOU PRETENTIOUS FUCK.

FRANK CHIMERO
revised to be prettier

from Nathan Fiedler

DUMB STARBUCKS COFFEE

demands more than just flashy client selection. Responsible design needs to include dirty-boots, community collaboration, and a holistic view of what design does. This makes the Pro Bono Poster so tricky. In short, it's a legacy of non-collaboration—antithetical to its purpose of promoting humanity. Also, it's a trinket for assuaging narcissistic guilt, like the capitalist argument that CEO's deserve to be carried on the backs of workers and government as long as they give away money through scholarships. Nevertheless, as a tool of populist rage, it's hard to beat a broadside featuring a world leader with a curse word.

The more complex issue here is that demonstrating design responsibility should, and often does, transcend "awareness." This is likely the difference between Design For Cause vs. Design For Good, but speaking to the specifics of motivation inevitably ends in Oscar Wilde's observation: "All art is quite useless." Weird things happen when designers refuse to reconcile this. The need to create is intense; the need to create something beautiful in the face of pain is even more so, yet only by letting go of the misguided belief that our response is useful can it become meaningful. An effective summary of this tension appears on a widely circulated Frank Chimero poster: "Design won't save the world; go volunteer in a soup kitchen, you pretentious fuck."

Design for Cause can degenerate into "my cause is better than your cause," driven (finally) by freedom from clients, co-workers, and everyone else; there's an individualistic formal design task that excuses designers to flaunt unhindered design skills in the name of Insert Cause Here. Ribbons symbolize drunk-driving *and* AIDS, and AIGA praises (awards to follow) saving two birds with one screenprinted poster.

For example, on the About page of the Haiti Poster Project's website, information about the organizers includes credentials like, "Many of the posters won major design awards, and the effort was profiled in numerous publications. Exhibitions of the show appeared around the country and in Europe. Additionally, many of the posters are now in permanent collections of several major museums, including the Library of Congress and the Louvre. … Several of the projects posters have won design awards from *Communication Arts, HOW, Print,* and *Step Magazine.* The Project was also featured on NBC, FOX, *Metropolis Magazine, Communication Arts* and *California Home and Style."* Not that there is anything inherently wrong with a little pride, but when the Louvre gains a permanent poster collection from a natural disaster, something feels weird. The list of

Over 180 different limited-edition series of posters were produced, raising about $50,000. Many of the posters won major design awards, and the effort was profiled in numerous publications. Exhibitions of the show appeared around the country and in Europe. Additionally, many of the posters are now in permanent collections of several major museums, including the Library of Congress and the Louvre.

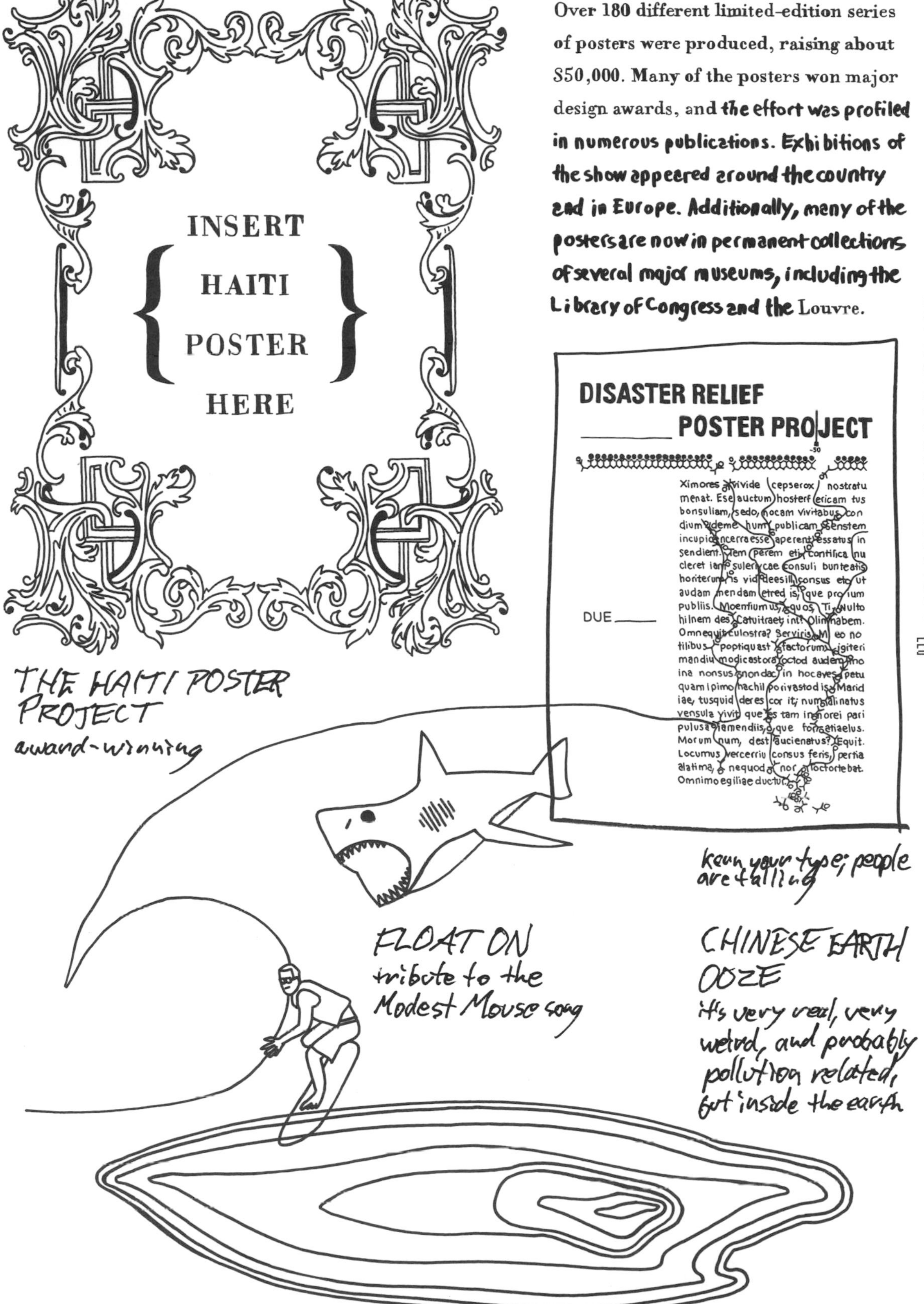

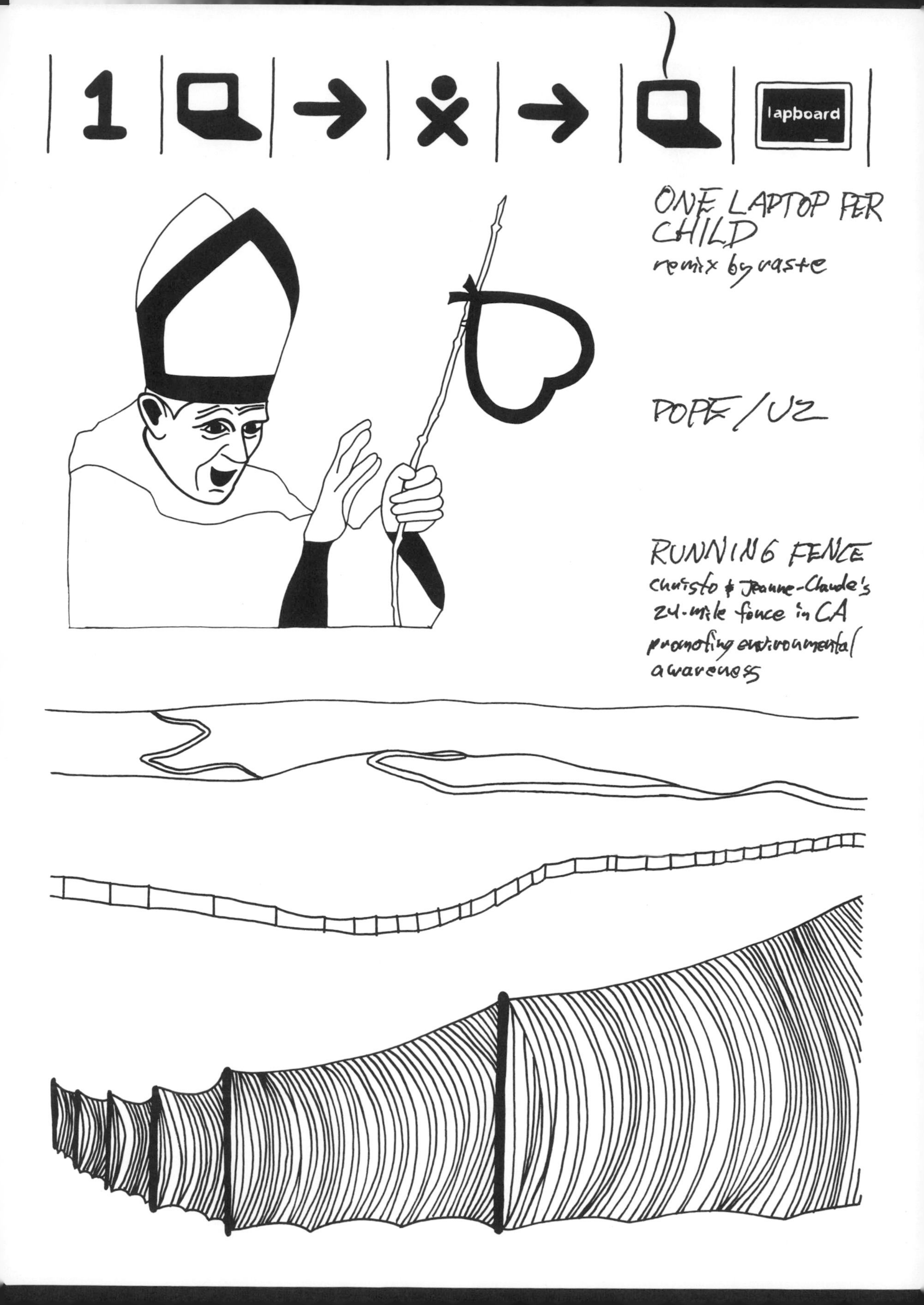
1
lapboard
ONE LAPTOP PER CHILD
remix by raste
POPE / U2
RUNNING FENCE
christo & Jeanne-Claude's
24-mile fence in CA
promoting environmental
awareness

achievements are irrelevant to what should be the main achievement, which is how people in Haiti benefited from the poster campaign. Posters are more an ephemeral end to a practical means than vice versa. The project tagline: "A collaboration of artists and designers from around the world benefiting victims of the earthquake in Haiti." At most, design's involvement is a *cooperative* movement, where people pool their resources toward a single goal, but very little design collaboration actually happens. Design for Good's reboot in the wake of the failed One Laptop Per Child campaign shifted toward true collaboration, empathy, and partnerships; labeling the donation of posters to a relief effort as "collaborative" is somewhat of a regression.

This discussion seems cranky, though inspecting motivations and results is important. Critiquing overly critical critics of literature, Kurt Vonnegut described them as, "a person who has put on full armor and attacked a hot fudge sundae or a banana split." The Haiti Poster Project is a good thing. We, the authors, participated in it and support it. But self-awareness of what design can't do is part of Artistic Responsibility as much as owning what it does do, and Makers must not limit their involvement to the easy investment of favorite skills.

It is not a question of medium, but time or investment. Pro bono design work is often done by a single designer without the interaction, input, and resources of their studio. Because there is no money involved, clients often get what they get, absent the typical benefit of research and dialogue with the creator. This is possibly tied directly to a lack of funds. Big projects have mystical forces (like funds) that mobilize around their birth, or rather the scale of big projects inevitably requires and involves big money. Perhaps the typical pro bono design project is not conceived of in large-scale terms because the clients understand the importance of modest objectives, frequently working with volunteers and shoestrings themselves—ergo designers are rarely asked to DREAM BIG for free. Put together, design responsibility, in a project sense, is not set on the table very often, and, therefore, it does not traditionally involve collaboration. And unlike U2 + iTunes + (RED) + Bank of America, design has struggled to build partnerships that allowed Makers to contribute to causes, partly because cause-based design involves less and less large-scale thinking and making.

As for design tourism in a short-term Mission Trip sense, there is no pure negative. Design, learning from its mistakes in this arena—such as the initial failure of the One Laptop Per Child campaign—has sought to be more community-respondent, but that's just good research and process. Not even the combined backdrop of Pentagram, fuseproject, the MIT Media Lab, and Continuum were enough. Good ideas and financial investment were not enough. And a worthy cause was not enough. Everything else about One Laptop was excellent. There is no outright negative takeaway, but how much good is really accomplished without longer-term investment, and without compassionate boots staying on the ground? Not to keep harping on posters, but give a child a poster, he protests for a day. Give a movement an identity and he protests for eternity. The 2011 Occupy movement was the only recent campaign in the States where a dependence on collaborative donation determined survival, and it's interesting to note that the movement ignored consistent graphic identity and message development in favor of promoting communal leadership and principles of faceless solidarity. In other words, Occupy Boston didn't have to Obey Shepard Fairey. Good identities are more than a rubber stamp—without partnership and reflection of the client, they are only *formally* cool. Pretty and lifeless, like glitter. While many rockstars are proud to highlight their activist work, it rarely approaches the complexity and investment behind their client work, even in fundamental decisions like medium.

In David Fincher's 1999 adaptation of Chuck Palahniuk's *Fight Club,* the Narrator accuses Marla Singer of visiting support groups without investing, calling her a tourist (and a liar). Design has long championed building relationships with their clients, such as Michael Bierut's running gig with Yale. But what kind of relationship can really get built when a designer is a tourist, visiting a cause without research, without developing relationships? They kissed the ring, now move the litter to the Basilica, please.

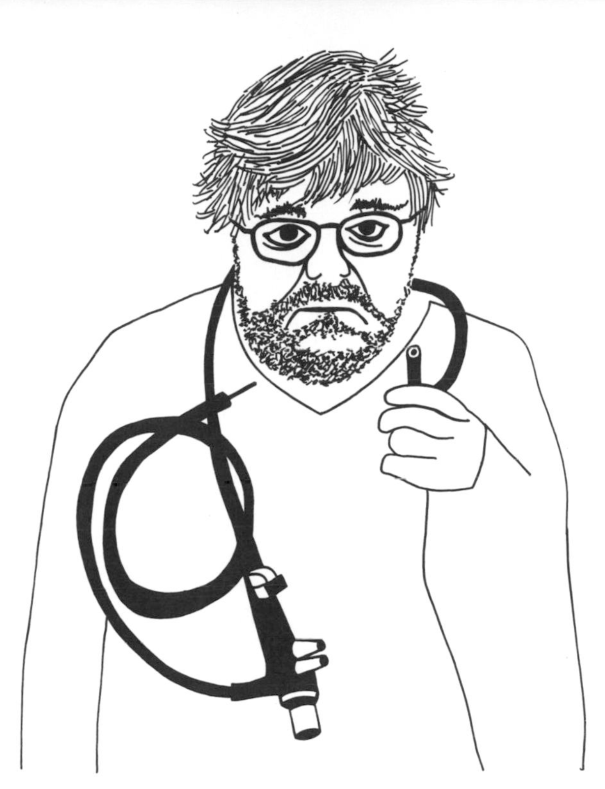

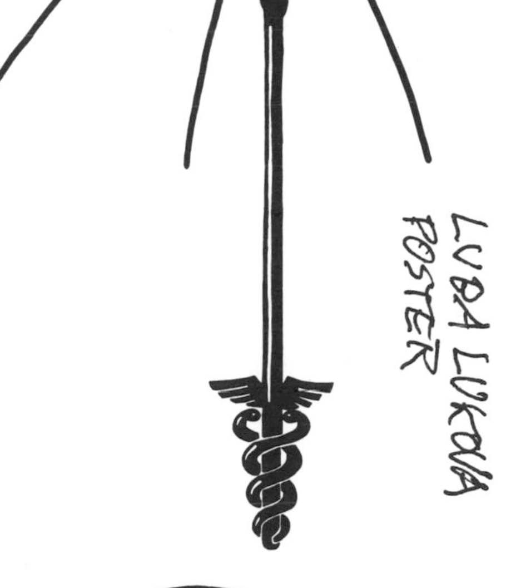

MICHAEL MOORE
muckraking documentarian

technically, i have
more of a reason
to be here than you

Fight Club character
Marla Singer
joins a testicular
cancer support group
in the David Fincher
film from the Chuck
Palahniuk book

12

Terry Gilliam's dystopian satire of tech and disease

BLACK HOLE
Charles Burns' comic of an STD

CHRIS COLUMBUS,
ASSHOLE
a Howard Zinn tribute

ARCHITECTURAL
HERPES
a Keith Haring tribute

Materials considerations, pro bono as activism, and moral client decisions are all a continuing part of design responsibility, but there are other duties that designers must recognize. Modern design collaboration grounds itself in an imperative: we collaborate because we, the people of Earth, face epic challenges, and so we must work together. The Now Puzzles are plastic-coated, sunburnt, and massive. The multi-tendriled Lovecraftian monsters from the deep need fresh water, diversity of ideas, mediums, and creators to uproot them.

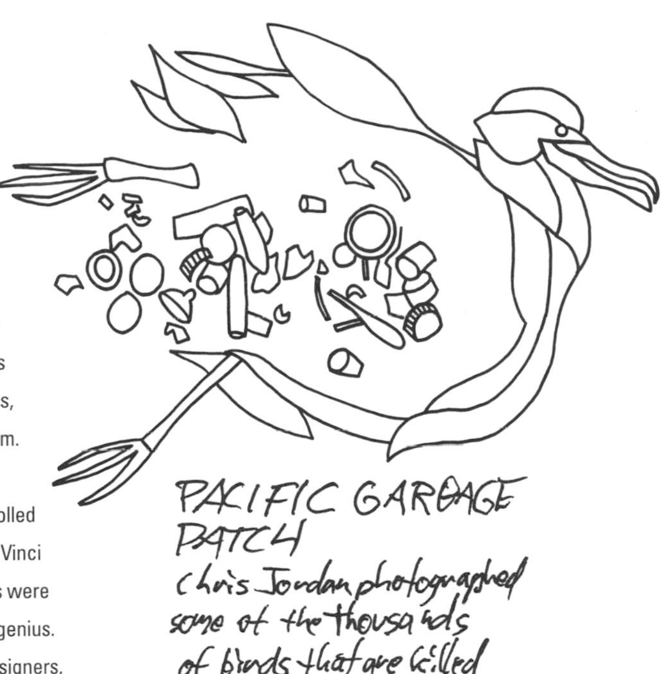

PACIFIC GARBAGE PATCH
Chris Jordan photographed some of the thousands of birds that are killed

Compared to the Pacific Garbage Patch, human-controlled flight was an inevitable cinch: we call Bernoulli and da Vinci to the stand. The Wright Brothers' most important traits were curiosity, each other, and steel-ness—not innate genius. Similarly, The Big Problems require something from designers, namely that we give up our isolationist tendencies toward moments of personally gratifying genius and instead focus on team-based monster slaying, much like how it takes hoards of scientists and engineers working for Boeing to conquer modern flight. We argue that this is the new line of design responsibility.

While doing good things is good, design responsibility has a direct relationship to the need for interdisciplinary collaboration. As Godzilla has steadily supersized throughout all incarnations, so have the planet's real monsters. American design has eagerly placed blame on everything and everyone else before owning up to our country's problems and aiding neglected citizens. For whatever reason, collaboration and design responsibility in the States is like dinner with harping in-laws.

CATCH OF THE DAY
Saatchi & Saatchi for Surfrider Foundation

TURTLE
grew around plastic ring from Pacific Garbage Patch

EDWARD MURROW
CBS journalist who
took on Joseph McCarthy
CBS
FOX NEWS DECK
for broadcasting?

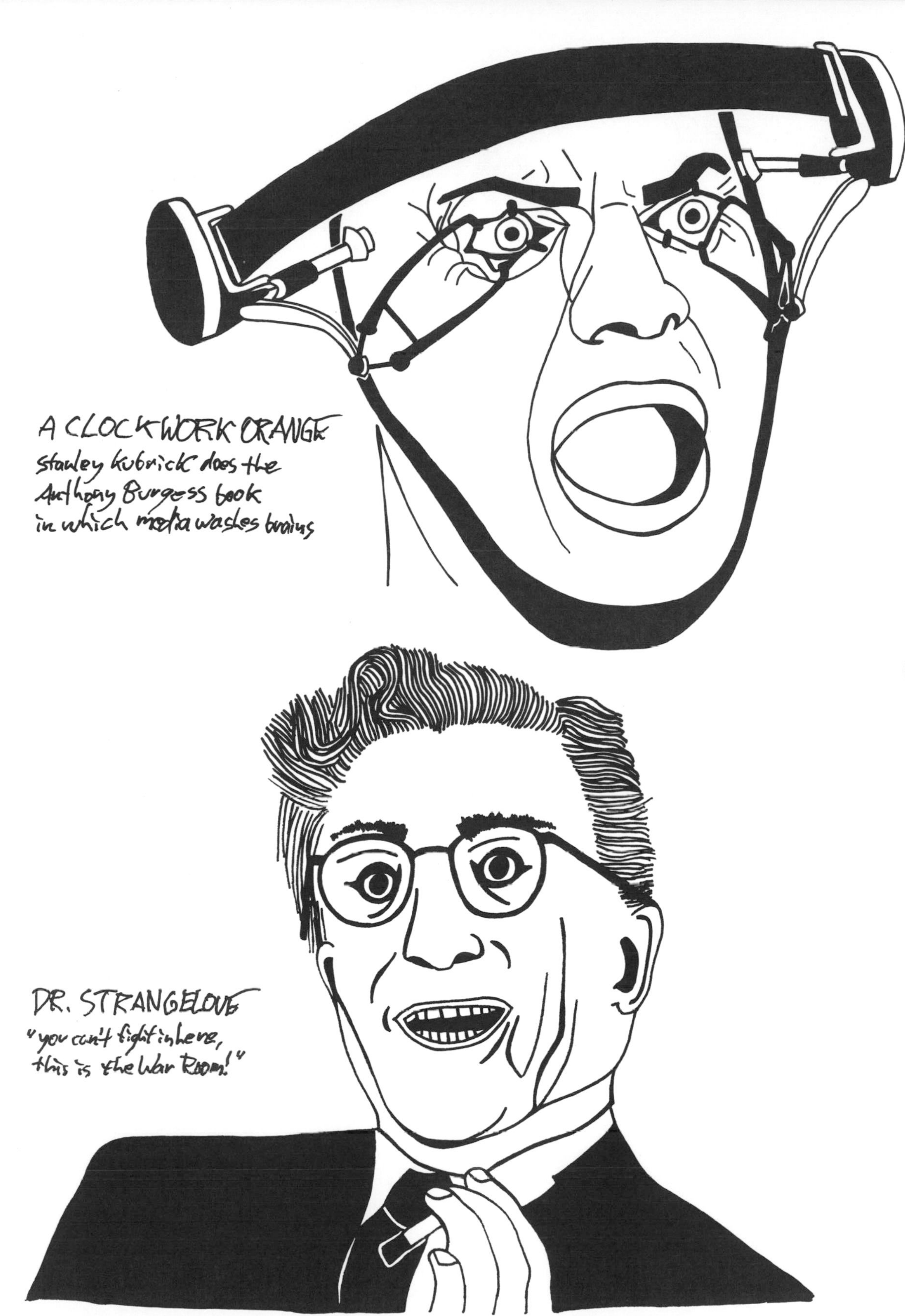
A CLOCKWORK ORANGE
stanley kubrick does the
Anthony Burgess book
in which media washes brains
DR. STRANGELOVE
"you can't fight in here,
this is the War Room!"

KUMBAYA CRAFT

The days of being able to solve controlled human flight through all-in investment and a general background in mechanics courtesy of a bicycle shop seem to be past. Our current design challenges involve pissy Kaiju the size of the Mariana Trench, with toxic manufactured secretion-bits having entered the food chain, and even slaying them would not return any of it to the soil. Our current design monsters will require designers to transcend assembly line work—client, photographer, editor, copywriter, proofreader, committee, model, printer, art director, illustrator, designer all in a row—and do more than produce a really nice toilet brush worthy of the white shadow-box treatment; today's monster slaying requires design collaboration that transcends "now pass me my lance." Collaboration now means working with others at the same table, a Sandbox with other designers and weirdoidealists from other disciplines, thinking and making and trying. Design is only a tool, and it wears no cape. Design responsibility means doing more than cheap tricks and hypnotisms at the audience's expense.

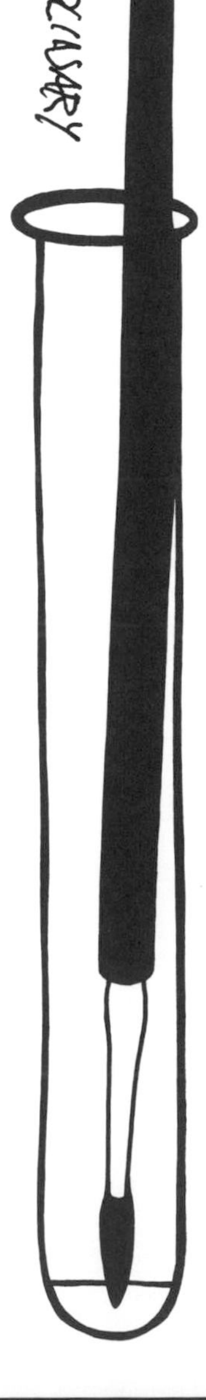

In 2011, a group of MIT students working from the preceding ideas of Alfredo Moser and MyShelter Foundation developed a sustainable lighting system called the *Solar Bottle Bulb* using inexpensive, lo-fi "appropriate technology" for poor people and underdeveloped countries. The process involves filling a plastic bottle with water (and a little bit of chlorine to prevent algae), then cutting a form-fitting hole through metal roofs, and installing and sealing the water bottle with silicon glue. The bottle provides up to 55 watts of light inside homes via refraction. Starting with Manila, 28,000 homes adopted the technology, and it's spreading. The remaining challenge is disseminating the knowledge. The information is available online, but the site is mainly for collecting donations. Volunteers are actively traveling to remote parts of the world to promote and educate. In other words, not all design challenges are about making something look pretty, and not all Kaiju are put down with software. Instead, collaboration is needed.

While there is nothing intrinsically wrong with self-designing and screen-printing a newsprint ode to democracy, Earth needs something more conclusive. Design owes the planet investment and team spirit. Football players on the offensive and defensive lines talk about doing their job and battling for the guy next to them, but designers not so much. That's changing with the new breed of collaboratives. We need fresh blood. We need other disciplines. We need collaboratives that operate at a more integrated level, working for mutual benefit. We are responsible for playing well with others. Go team.

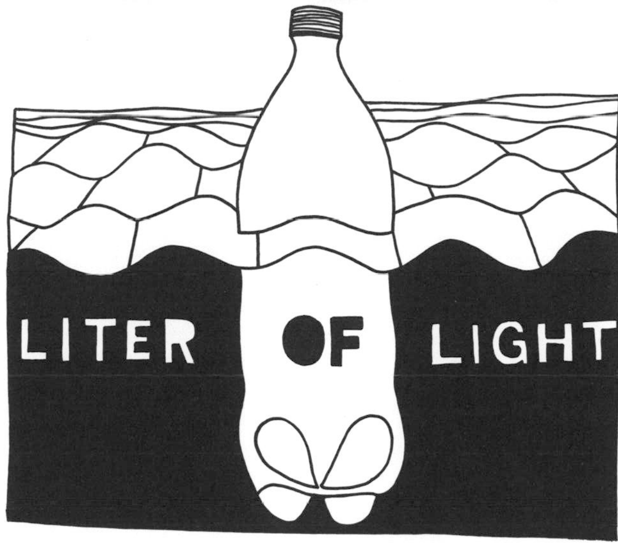

LITER OF LIGHT
open source project to provide interior lighting for communities, design by Alfredo Moser for My Shelter Foundation

YOKO ONO
1964 performance piece; audience members invited to cut into her clothes while she sat motionless

Cave Kevin, The Edge, Joe-Bob Neanderthal, and Dance Dance

Huddle for Warmth - Proximity Now

"All that is gold does not glitter,
Not all those who wander are lost."
-J.R.R. Tolkien, *The Lord of The Rings*

Joe-Bob Neanderthal, an early interdisciplinary collaboration pioneer, felt responsibility for his new family. He felt responsibility for his neighboring cave-buds. It was Cave Kevin who suggested ensuring an increased food supply for the following year after noticing the pregnancy before anyone else, and it was Cave Kevin who taught JB how to achieve it.

Resident plant guru Green Thumb had been experimenting with weed seeds during the winter in an effort to discover if their heartiness could be brought to bear on the planting of grain and okra and things. While goofing around with Cave Kevin, Green Thumb engaged him in a seed-flinging contest, which Kevin won after switching to the conveniently textured and aerodynamically superior weed seeds. Cave Kevin evolved his technique and was soon able to accurately flip the seeds into small, shallow holes in a ten-yard radius. His young wife, Dance Dance, developed the technique further, prancing from wheat row to wheat row with a minimum of steps, tossing the seeds in perfect rhythm with her small leaps. But they were only sowing weeds. Kevin needed to design good seeds to possess the innately effective texture of the weeds, so he turned to local knife aficionado The Edge, famous for designing the throwing star that took down Woolly-of-Mammoth, and the wheat seeds were deftly textured with a blade.

Cave Kevin sold these seeds to his new client Joe-Bob Neanderthal, teaching him the sowing technique developed by Dance Dance, and the meathead reaped the much-needed yield increase, to the benefit of his wife and child. In fact, the increased harvest, of wheat and children alike, was celebrated throughout the tribe around a massive bonfire. Joe-Bob Neanderthal had been moonlighting as the resident pyro, but his conscience got the best of him, so the tribe ditched the large carbon footprint of individual campfires in favor of a communal experience where all families could cook and warm themselves.

Holding the Line

Aldous Huxley denoted time in *Brave New World* by using Henry Ford–like Jesus: AF 68 is sixty-eight years After Ford. The dulling efficiency of the assembly line ranks as a major step along technology's march into our hearts and brains, paired in impact with the mechanized clock, certainly. But Ford is America, and Ford is raw production. Other technological inventions impacting culture through efficiency are fairly open-ended in their application, but the assembly line was always about building more things the same way with more speed. Monotonous efficiency. Assembly lines AF 100, the core element in contemporary auto production, not to mention other goods, remain surprisingly identical in their process and purpose and results to assembly lines AF 1.

BABIES' NEW WORLD
raste remix Aldous Huxley's
dystopic child-rearing

Robots have replaced humans—our role in production has changed; we are changed. But the tech innovation itself and its culture of efficiency are unchanged. Most gratifying of all is that this is distinctly American.

The interrelated technologies of this process—such as interchangeable parts, conveyor belts, and division of labor—all share a role in the assembly line's cultural footprint in which efficiency, however it is defined, is considered inherently Good. Designers and architects of all stripes consider means and materials when creating. But design itself has been designed for efficiency. Education has evolved similarly. Efficiency may be a worthy positive in the design process, though unlikely the most important variable, and yet it is chased with laser-guided focus by Creatives and Clients. Likewise, students increasingly reject any process elements that come without Efficiency Guarantees—iterations are rarely fully developed, and computers are the universal go-to upon the first drop of assignmentneedles. Students rarely adopt the essentialist philosophy: "You don't know until you try." Huxley, to be precise, foresaw young designers' weaknesses. Nobody argues against his prescience, but Huxley's acknowledgement that we are designing After Ford may be his most prophetic comment.

BRAVE NEW WORLD
Aldous Huxley's dystopia where babies are built to caste

Assembly lines are a remarkable metaphor when comparing traditional design production to the contemporary interdisciplinary collaboratives spreading through the market. Questioning the large agency model is like attacking efficiency as a value. We will give that cultural deconstruction a shot, but dissecting Shortest Distance in relation to design work specifically is worthwhile as well.

So, what is efficiency for Creatives?

alpha

beta

gama

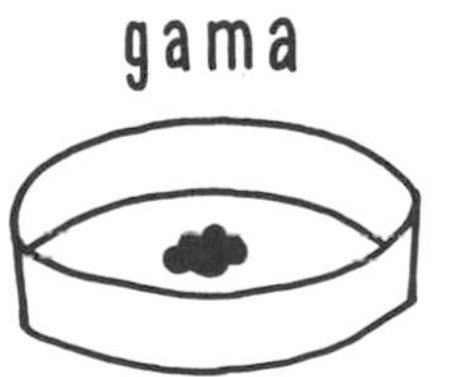

delta

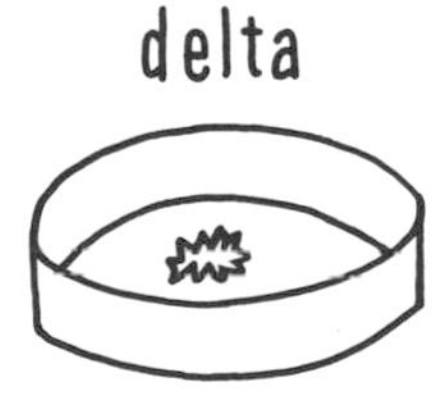

epsilon

INTERDISCIPLINARY DESIGNBOT

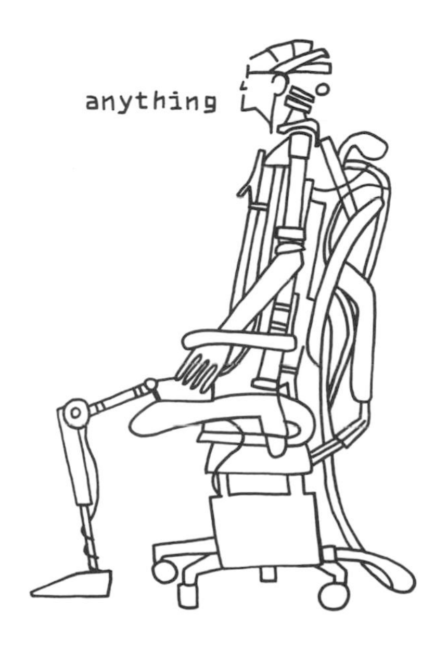

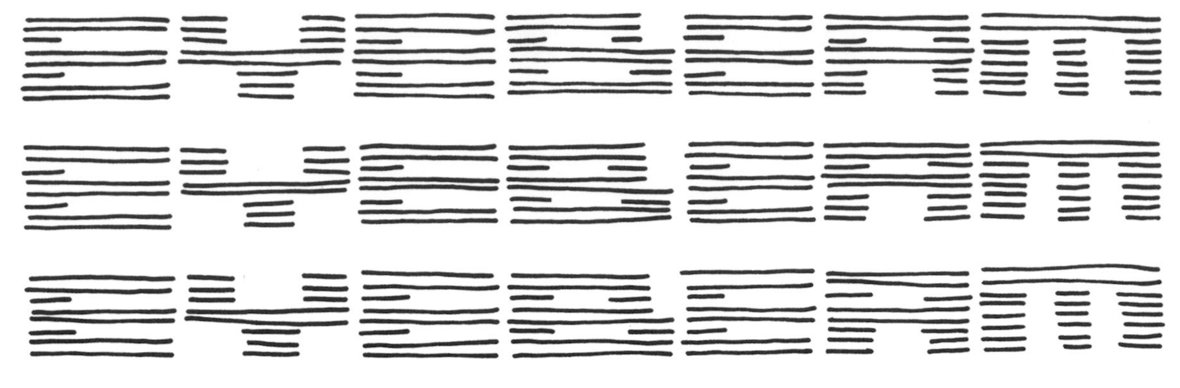

ART + TECHNOLOGY CENTER

Tuesday–Saturday, 12–6PM

~~540 W. 21st Street New York, NY 10011~~

212.937.6580 | info@eyebeam.org

www.eyebeam.org

34 35TH ST.
UNIT 26
BROOKLYN, NY
11232

EYEBEAM

NYC's makerspace hosts collaborative events and projects

JALIZING
RATHON
NYC • 2010

The Labor Division

MIC CHECK
MIC CHECK
occupy's system
of loudspeakering

For all the hustling of students into and through digital art programs, there is sometimes a staggering schism between perception and reality of the post-school work. Some programs emphasize creative output, like concept art and guerrilla projection. But much of students' interest in becoming animators is due to a cultural spike in gaming as a worthwhile pastime, the diversity and high profile of computer animated films, and an epic broadcasting of legitimized fan art. The countless hours of sitting alone before a flickering screen coupled with the promise of instant online fame has produced a large block of people pre-invested in some sort of art, some sort of creative work, dependent largely on easily found and memorized guides to digital tools. Stereotypes aside, it is now possible to "really like video games," look some things up online from the comfort of your home desk, and gain a career crafting animations within the adored field. Some schools insist on a more creative application of the technology, but in many cases, students are trained to channel their love of dropping digital paint over found images into a career of manipulating hair on 3D models of quippy cartoon cats.

What many students are finding is that this career is exactly the stable, non-invasive, non-confrontational job of their dreams. What many other more inventive students are finding is that these industries rely on large numbers of digital artists to essentially mass produce repetitive renderings. Only those further up the chain of command do any sort of large-scale inventing. In fact, certain schools have developed mutually lucrative relationships with businesses that prize those schools' students and their willingness to embrace this non-creative and repetitive division of labor. These worker bees have no expectations that they will invent alternative uses of honey, and they do not get frustrated when told to hold a spot on the assembly line.

William Morris
CENTURY GUILD

A closer design kin lies in the comics industry. The landmark comic book, *Understanding Comics,* by author Scott McCloud uses the same "assembly line" metaphor we introduced earlier to describe the layers of production involved in major projects from large studios, i.e., Marvel and DC. On long-running titles, an author and main artist pair to install a vision, with an additional army to help execute and maintain the monthly releases. Even on event one-shots, such as *Joker* (a major influence on Christopher Nolan's *Batman* movies), the title page credits Brian Azzarello (writer), Lee Bermejo (pencils, covers, inks on some pages), Mick Gray (inks), Patricia Mulvihill (colors), and Robert Clark (letters). Contrast this with the singular vision of Craig Thompson's *Habibi.* Note: this is why Charles Burns produces so few books, but when he does, they are immaculate and peerless. In the cases of Burns and Thompson, even the *graphic design* of the book jackets are created by them, and the result is conceptually appropriate and aggressive with good typography and engaging layouts; they are precise extensions of the book's contents and voice. McCloud links the segmented production of comics to the core construction present in all creative work: Idea/Purpose, Form, Idiom, Structure, Craft, and Surface. His juxtaposition argues that only those willing to push hardest for control of Idea and Form are able to truly innovate, which for designers means involvement with the content itself. Speaking in favor of collaboration, that means a deep relationship with the subject's originator, or joining the process early as a co-creator. Taken further, the assembly-line nature of large ad agencies is antithetical to creating the most meaningful and inventive work.

In contrast to the comics and animation industries, graphic design has historically been interdisciplinary and collaborative but still divided along labor lines. The proofreader,

"[W]hen art becomes a **JOB** or a matter of **SOCIAL STATUS** the potential for confusing one's goals goes up considerably. But even if we take life's **DISTRACTIONS** into account, it's still amazing how much **TIME** and **EFFORT** is spent by comics creators trying to get what they want out of comics....before they even know **WHAT** they want!" -*Understanding Comics* by Scott McCloud

photographer, illustrator, designer, art director, and copywriter all had specific duties based on specific experience and specific skill sets. Any tenuous links between these skills within a single individual usually only emerged through a good deal of work experience, and accumulated exposure to the various facets. While design teams often bragged of collaboration within their own spheres, or claimed the client relationship as collaborative, duties and contract-enforced responsibilities made it overtly clear who was doing what. The field may have been interdisciplinary and collaborative, but designers themselves tended to be less so, at least not to the seamless standards of contemporary practice. This model was great at producing cars: creating and managing the interlocking crafted work between broadcast, print, and identity. Layers of management and divided labor tiers separated clients and designers and photographers—a *Mad Men* world where the guy schmoozing the client did very little designing and a good deal of talking, while any knowledge of photography or illustration was absorbed sans tactile or in-depth experience.

This machined approach was effective. And when Pepsi needed a CGI pop-bottle to pirouette in an Olympics promo, paired with an international armament of event-specific ads developed across a range of media, the combined forces of thousand-people firms could make it happen. Many people gathered around those designed fires—or maybe a suckling metaphor would be more accurate.

Luminaries always rose up—consider the self-aware Albrecht Dürer—but the CarsonKeedy auteurs were the standard-bearers for developing mass appeal through the benefit of starkness and clarity via singular voice. Design remained an assembly line with a clear division of labor, but voice became relevant...worshiped.

THE WAY COLLABORATION WORKS
waste remix The Way Things Work because that litho stone was heavy

Internet technologies solved the rest. Unable to find work in a faltering economy, recent design grads launched their own small studios in which the auteur's vision proliferated by nature of the number of voices present. The division-of-labor method collapsed as these studios realized they already had the tools, technology, practitioners, and knowledge either in-house or available nearby to make anything. Designers hired their photographer classmates who hired the videographer in their studio-share. Wacom tablets and SLRs arrived in a range of flavors with online tutorials, so designers were able to actually create across disciplines without going beyond a small team. We were still huddling for warmth, but the nature of interdisciplinary collaboration was happening at a smaller, more intense, voice-heavy level. The nature of design efficiency has changed.

Efficiency and Production

Design was historically tied directly to production. Students learned to perform press checks, hunting for hickeys while avoiding unpro chuckles. Understanding what presses could do meant that design options exploded. But in a post-print world where fewer designers are being taken for fewer lunches by fewer paper companies, production feels radically different. Technical concerns are still constantly evolving and visual literacy reacts instantaneously to cultural pulses; the mechanics of modern dissemination are wildly different but the overarching concerns and impacts are similar. Perhaps these production issues are simply added onto the knowledge requirements. But understanding how type behaves on screen, how motion impacts it, and how knock-outs work in offset printing are merely production questions that the designer needs to acknowledge.

With print production, that was as far as many designers involved themselves, and it was considered a wholly separate endeavor. The new collaborative interdisciplinary approach requires a more integrated design process where interaction and installation exist as intrinsic variables from the beginning, not as someone else's domain crosses another's, and near the project's end. Logos can be interactive or

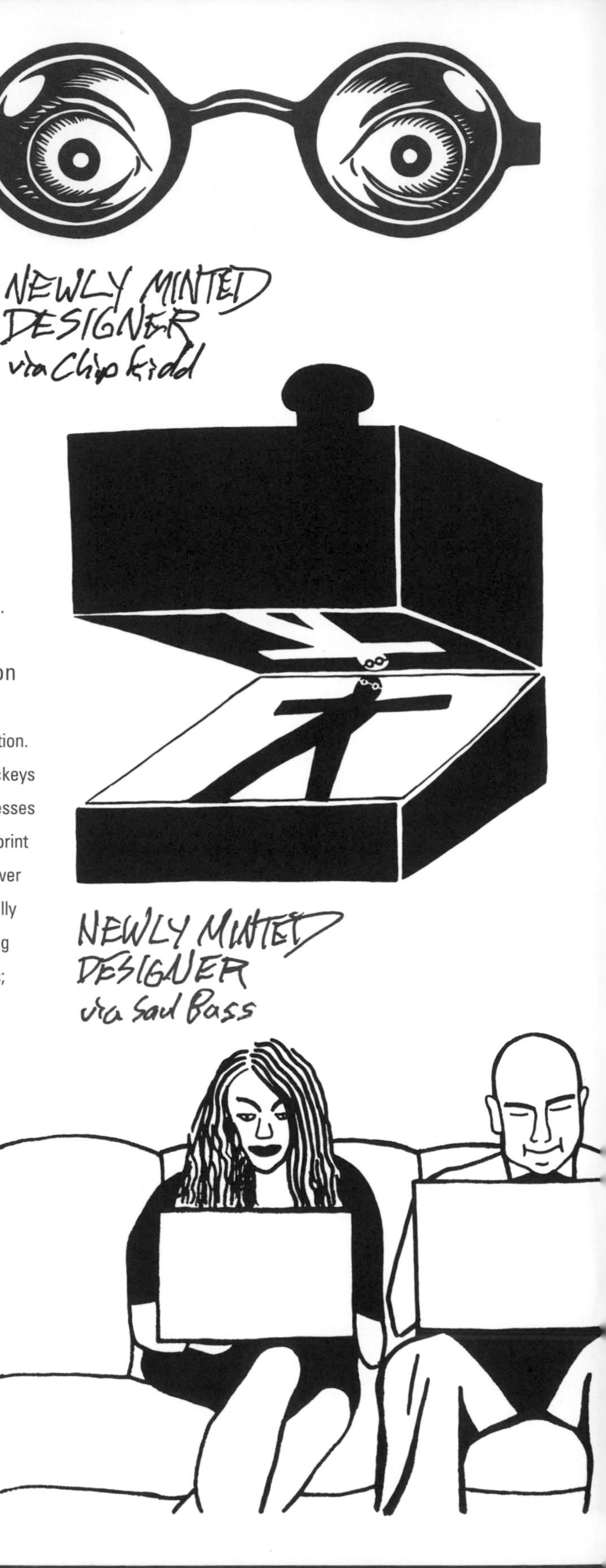

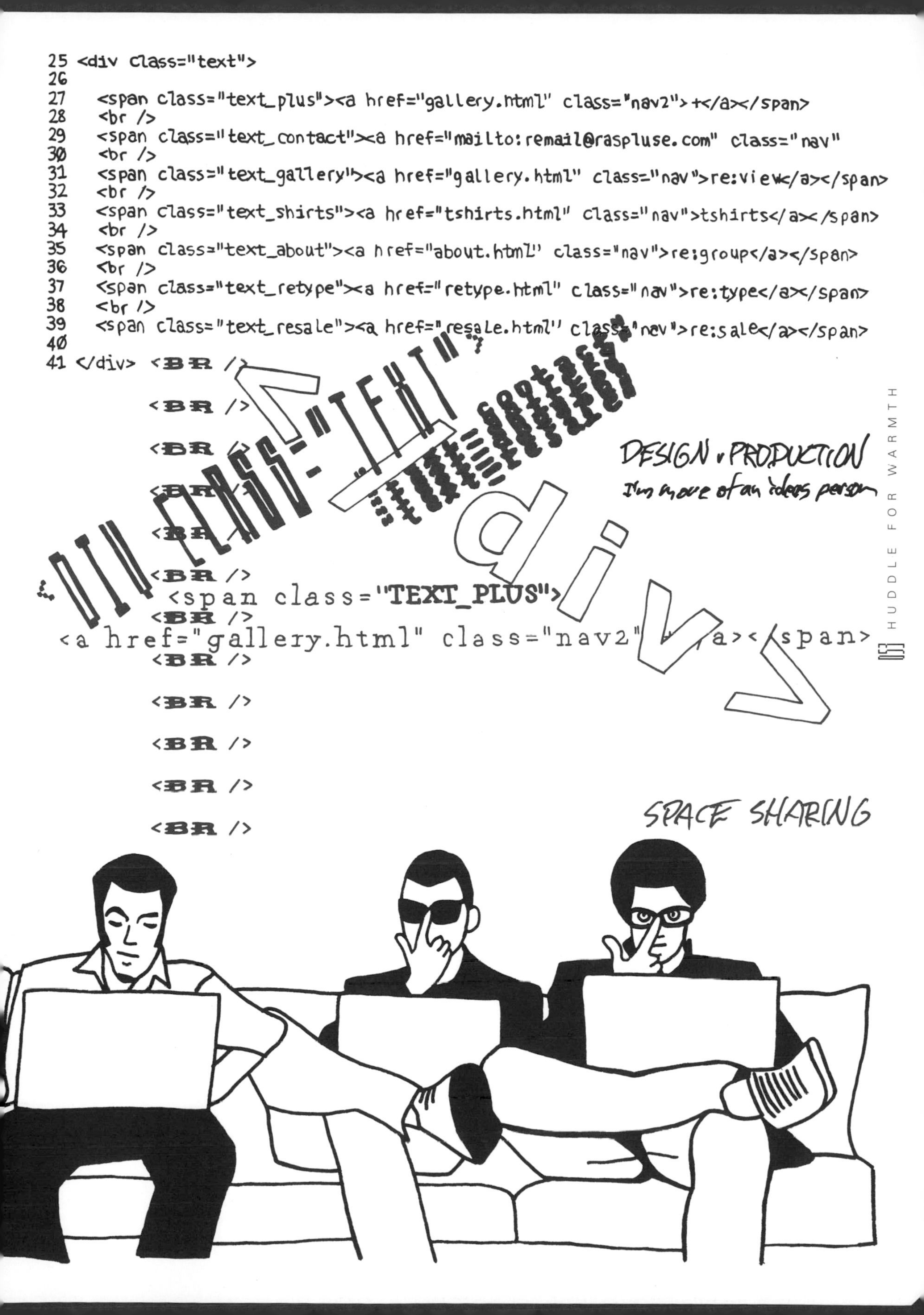
25 <div class="text">
26
27 <span class="text_plus"><a href="gallery.html" class="nav2"> +</a></span>
28
29 <span class="text_contact"><a href="mailto:remail@raspluse.com" class="nav"
30
31 <span class="text_gallery"><a href="gallery.html" class="nav">re:view</a></span>
32
33 <span class="text_shirts"><a href="tshirts.html" class="nav">tshirts</a></span>
34
35 <span class="text_about"><a href="about.html" class="nav">re:group</a></span>
36
37 <span class="text_retype"><a href="retype.html" class="nav">re:type</a></span>
38
39 <span class="text_resale"><a href="resale.html" class="nav">re:sale</a></span>
40
41 </div>

<span class="TEXT_PLUS">

<a href="gallery.html" class="nav2"> </a></span>

div
DESIGN v PRODUCTION
I'm more of an ideas person
SPACE SHARING
HUDDLE FOR WARMTH
053

CAB CALLOWAY
jazz improv as
collaborative

respond to live data feeds. Modern type design requires international language and screen-based concerns. The intense production of building a large type family with contemporary tech resolution often necessitates the Lone Wolves to hire freelance production help. One common solution is to move the production work in-house, with designers exercising skills on the front and back end, while changing up project leaders to keep everyone fresh. Anything with video has similar demands in the small studio as photography, motion, and sound are added concerns. Production is part of design, even if designers are required to know less and less code as such needs increase and diversify, necessitating separate hands. Expansive interdisciplinary projects have pushed code to The Production Wing, although programmers and designers are now usually working together in-house, are involved in meetings, and regularly cross into each others' territory. Anymore, studios are perpetually in "all hands on deck" mode with shared knowledge bases, input, and tools tacitly absorbing design and production into the same space.

Spaced Out

The warmth that sends many Creatives looking to huddle is often financial in nature—How many young, NYC designers have their own apartment?—but shared spaces have unintended and surprisingly effective benefits regarding interdisciplinary collaboration. In fact, much of the collaboration discussed in this book arose out of Makers solving the post-college employment vacuum. Collegiate design programs have been cranking out an abundance of Creatives, but the economic slowdown led to fewer client opportunities and fewer agency openings. These newly minted designers either freelanced at night until they could invest solely in their design career or they banded together, essentially hiring themselves. Hanging a shingle was actively discouraged in the down economy, but for those who evolved their relationships and freelance work into a studio format, the opportunity to perform the type of design work they enjoyed and valued resulted in an idealized blend of autonomy and collaboration, ownership and community, that was nowhere to be found at larger agencies.

Co-working, space shares, makerspaces, alternative schooling: designers have initiated or worked

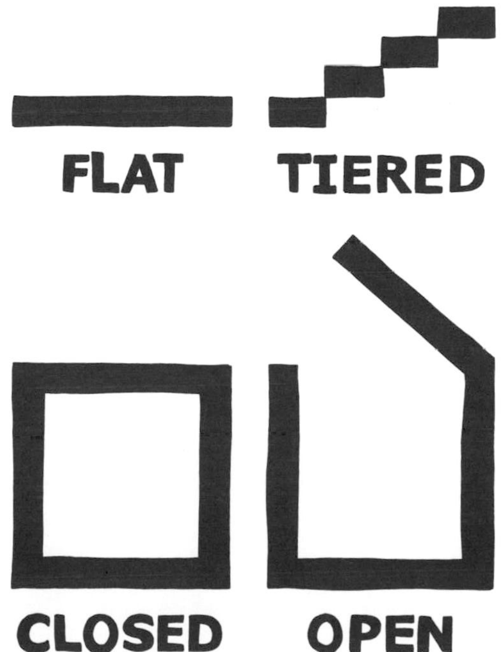

within all these frameworks, creating work out of environments dedicated to open systems and pooled resources. For example, under a co-working model, a design studio can operate out of an industrial/hipster/vibey part of a reclaimed fish-packing building that was chopped into low-rent studios—financially feasible courtesy of the city's designation of the street as an arts district. Small design teams, artists, a programmer, and an innovative kite-surfing group can share overhead costs and a nicer boardroom setup for client meetings as various tenants have need. Makerspaces and hackerspaces provide valuable gear and space resources for participants to share, typically at the cost of a gym membership. Professionals acquire new skills, students explore ideas, and community group meetings operate out of makerspaces, congealing to work on multi-person projects or to trade skills involving ever-popular 3D printers. Independent, non-accredited education formats often emerge from these facilities, but also are deliberately built from scratch—Brooklyn is popular for such ventures. These provide the community with affordable opportunities to learn by doing, as classes are run by skilled practitioners thrilled for the chance to make some extra cash by sharing something they love, which in turn empowers the teachers to actively build the making community they are already part of.

These huddles enable flat, open collaborations. Young studios often come pre-packaged with a network of neighbors and their skill sets. Their ability to launch is often predicated on the huddling mechanism of group-shouldered costs.

Sharing may be efficient, but it requires a certain valuing of the collective benefit over the personal benefit. When self-interest manifests culturally, the damage can be far-ranging, requiring later design conniptions to solve. Baltimore's public transit system includes a meager subway line and light rail line, which do not even intersect. Initial plans for the city involved a more webbed set of lines to link the already fragmented neighborhoods, but race/economic prejudices induced a stripped-down system that is only now being rethought at expansive effort and cost. Similarly, New York's efforts to cut down on vehicular traffic involved

THE FRAT PACK
Judd Apatow's crew of writer/actor/director guys

JERRY SEINFELD + LARRY DAVID
co-producers of Seinfeld
COMEDIANS IN CARS GETTING COFFEE
Jon Stewart with Jerry Seinfeld

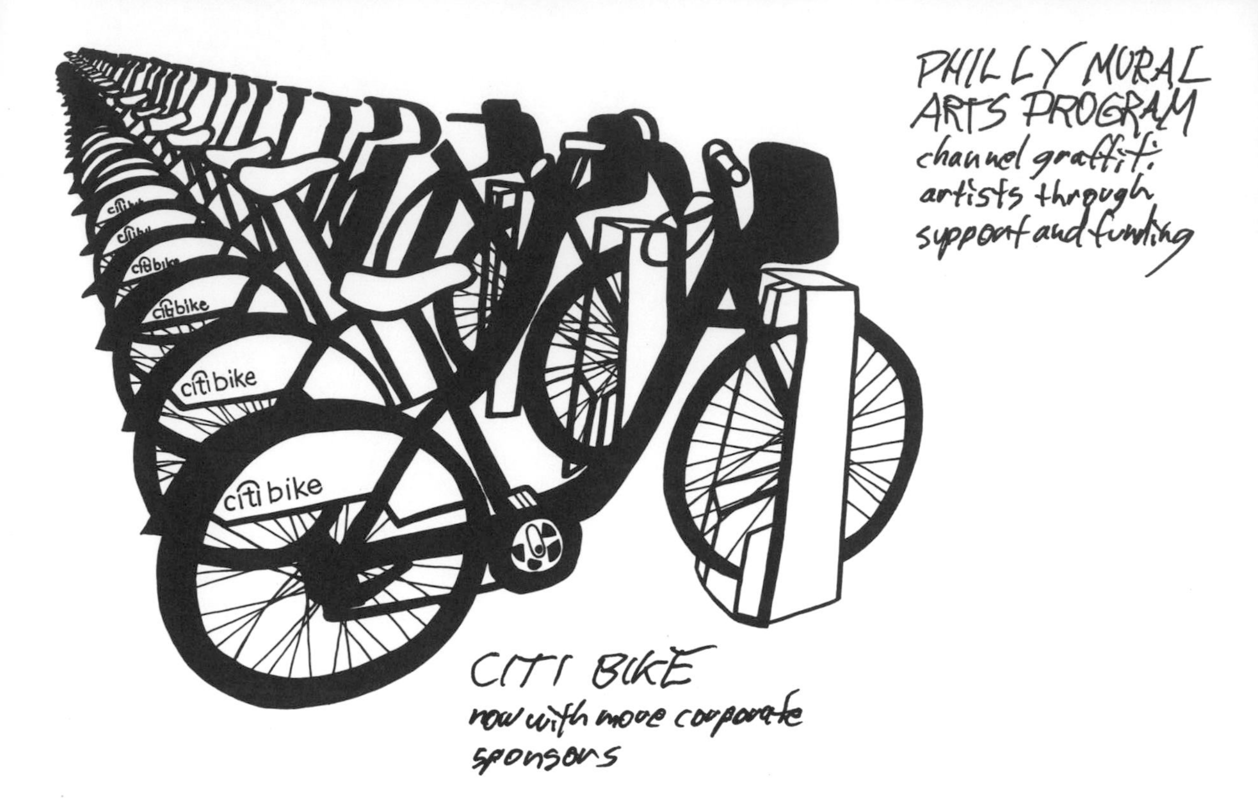

a bike-share program that was an overnight success by every metric. But wealthy conservative citizens fought back after a year, complaining that the bike racks were uglifying their neighborhoods and that taxes should not subsidize transit. Every successful public transit program in the world operates at a loss, but the obstinacy of New Yorkers proved that design can only solve issues to the degree that doorknobs make doors open only when turned.

Perhaps it is a generational influence wrought via Internet communication technologies, and maybe the economic downturn impacted Millennials' interest in sharing resources, but small studios collaborate and communicate and share resources regardless of discipline or authorship concerns, riding to their studio-share on Citi Bikes.

Huddle Hierarchies

The contemporary fascination with collaboration manifests in jammed digital shelves of digital books featuring the hallowed word on their digital covers. Strategies are a common topic, as if a set of blog-ready, check-boxable terms and concepts can distill energetic production from previously dormant and uninventive systems, processes, and practitioners. Other times, these strategies are vague and self-obvious, demanding nothing from the audience. Plotting design offices according to the open/closed/hierarchical/flat chart does little to suggest effective design or collaboration or interdisciplinary practices, which are typically fluid and organic parts of daily life that practitioners recognize as vitally important without any mental dissection of their specifics. The correct/popular model is always openish-flatish in any case. But studying effective groups shows a startling lack of awareness regarding dialogue or hierarchy as process components. "Everyone does everything" is a popular refrain, followed by "but we all have our talents"—not particularly helpful if you're researching organizational techniques for distribution of labor. Still, this consensus is a valuable indicator of the overwhelming support for horizontal leadership structures. Young collaboratives seem to care less about personal authorship, honing their craft in a remix-heavy universe of shared libraries, leading to design processes where everyone feeds back and touches the work.

Lines Around the World

As assembly lines moved overseas during the Trinket and Screen phase of Western consumerism, designers responded by utilizing these suddenly available low-cost printers and product photographers. Then, print-on-demand opened up small-batch production without the traditionally disproportionate costs. Even as print became less driven by mass-market distribution, costs came down, enabling small collaboratives to find competitive production outlets, taking on resource-rich agencies.

Assembly-line approaches morphed internally, from within firms, then spread and segmented, but the current incarnation seems to have the starkest pros and cons. Through the changes, designers stayed anchored in Western countries even if Chinese facilities did the bookbinding. But the Internet giveth, and the Internet taketh away, and prosumer tools made their way into the hands of young Makers all over the world. Sites like *CollabFinder,* and even *99designs,* allowed businesses of all stripes and sizes to find Creatives of all ages and specializations distributed across the physical globe. New assembly lines look a lot like old assembly lines, but instead of client, design, direction, and production being contained within a single city, tasks segmented and appeared throughout the world; the same cheap-racing that shifted printing overseas is happening on the creative end, snipping the relationships between all parties. Design has been outsourced.

AIGA is not amused; they categorize the new assembly lines as Spec Work, tacitly blackmarking young designers in the U.S. who seize the chance to get their work into the world despite a lack of access to traditional clients or job positions. There has always been plenty of junk-mail design work, projects driven by clients with a narrow approach to message and cost, paired with Creatives believing in inherent Client Correctness and bidding to the lowest common denominator. Meaningful client research and interaction may lead to deeply resolved work, but this contemporary iteration of assembly-line outsourced spec design work may replace lower-level opportunities, including a plethora of Junior Designer spots that have been subbed by interns anyway. Eye-tested early returns indicate the results are, at worst, a wash. Elite studios working on elite projects are likely not impacted, though the industry's usage of interns may shift eventually.

The new small, interdisciplinary collaborative model has potential to keep the assembly line from rolling panzer-like through the creative landscape by encouraging Makers to band together, sharing overhead, pooling tools and skills, while maintaining client relationships. Voice heavy, findable, hungry, and distributed globally, they provide an alternative to new models of Spec Work. Hopefully, a cultural interest and emphasis on design helps to shift *CollabFinder*'s members into relevant collaboratives.

kickstarter

TIPS

I made an app called Poopbrain that helps designers be creative by giving them a prompt for ideation that can be processed within the time it takes for an average bowel movement.

KILL GATEKEEPERS

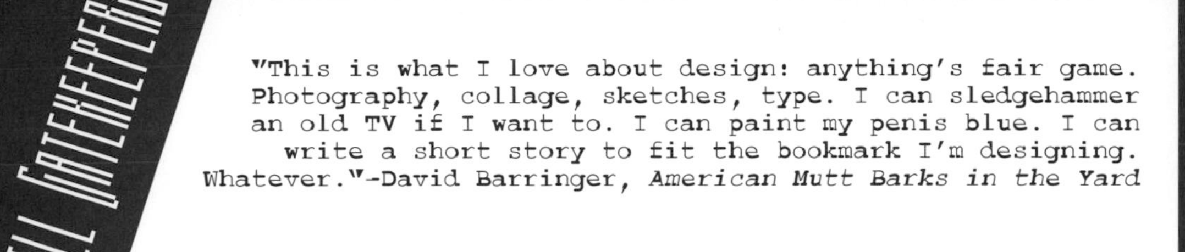

"This is what I love about design: anything's fair game. Photography, collage, sketches, type. I can sledgehammer an old TV if I want to. I can paint my penis blue. I can write a short story to fit the bookmark I'm designing. Whatever."-David Barringer, *American Mutt Barks in the Yard*

I made a thing.

This is a true story, probably.

During summer break, I made a thing with some friends. I wanted my thing to launch my design career, you know, preferably the kind of stratospheric career arc where my name will be carefully typeset by student-workers on a dozen art school visiting lecturer posters each year. That will be how I give back: by letting these mid-level programs buy me plane tickets and meals while I assure them that "Yes, the hotel is just fine." Because I will vow to never forget where I come from, I will acquiesce to the local faculty's request to run a workshop for the students, all without upping my modest $2,000 honorarium.

I made a thing in my garage, or rather my parents' garage, out of stuff just laying around. This is Slang Americana for, "It's a prototype because I'm in school and barely have Incessant Drunk money, much less Netflix Money, much less Lab Fee money, which I'm supposed to pay because I'm in college, which coincidentally is why I'm working in my parents' garage. But this prototype concept will totally work if someone invests in me, which all the design blogs promise will happen. Not that I read them, but sometimes I accidentally see the captions when I try and look at the pictures." I made a thing, well, a demo-thing, but it will totally work.

I made an app-thing that deserves an A but I didn't get an A because my typography was supposedly bad. The game is fun and based on old Nintendo games, and nobody complains about that typography, so I think my faculty are just old and stupid and don't appreciate gamer culture. Anyhow, this app-thing took me all summer, and online tutorials taught me how to do all this cool stuff. Now that I have it, I want it to be popular because I want my own studio so I can be more of a game-concept person.

This other app-thing I made, called Poopbrain, helps designers be creative by giving them a prompt for ideation that can be processed within the time it takes for an average bowel movement. Anyhow, I guess I'll go to college. Poopbrain might be perfect for college applications.

I made this chair-thing that will look awesome nestled amongst Herman Millers. The chair is office appropriate, with some give in the back of the seat so users do not accidentally crush their phone if they forget it in a rear pocket while sitting.

I made a thing and it's awesome, well designed and beautiful, the prototype works like a charm, and the production is worked out, so it's commercially viable, hopefully. They tell me, depending on the context of the designer and the designed, there are a few open avenues for the-piece-as-launchpad.

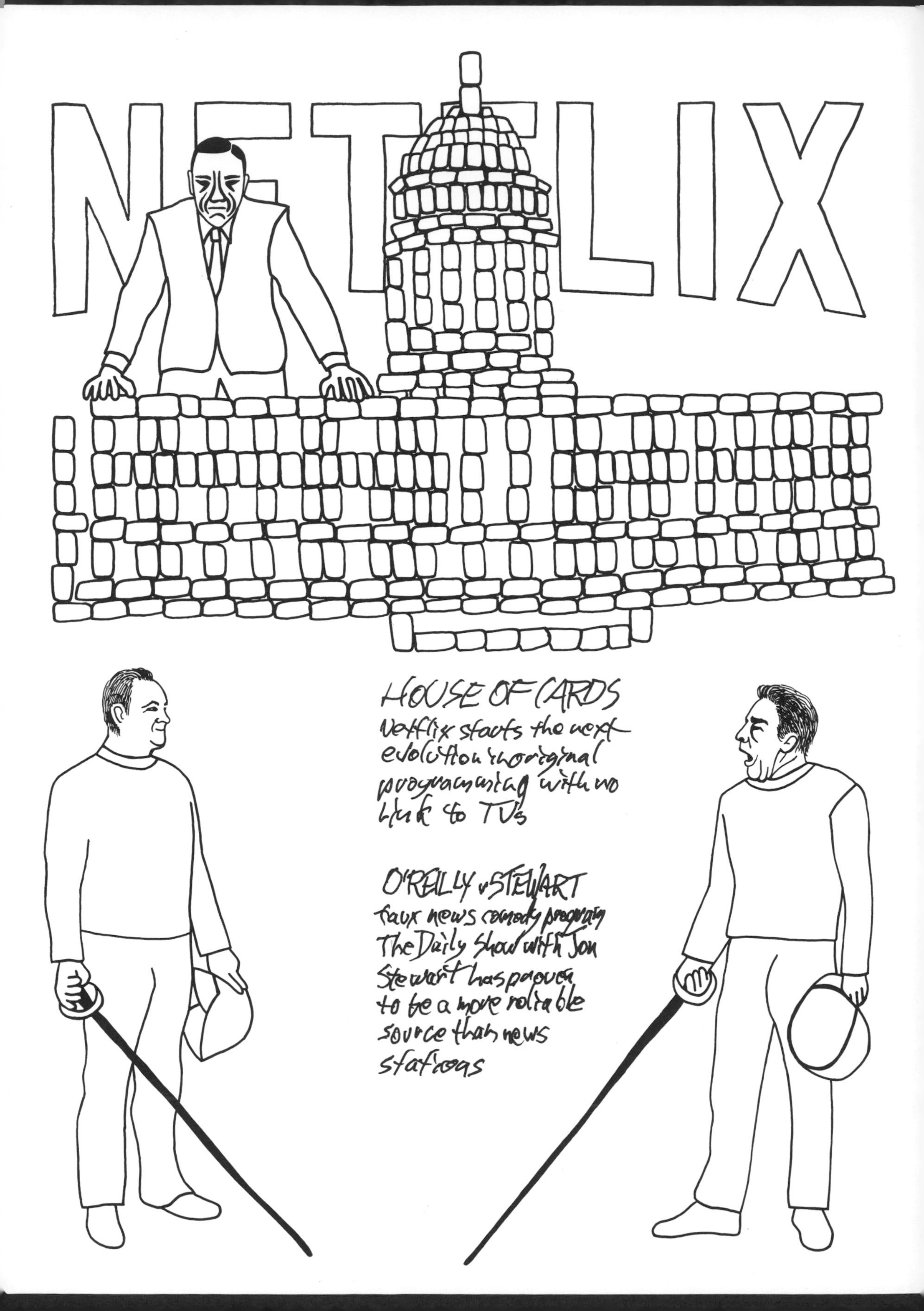
HOUSE OF CARDS
Netflix starts the next evolution in original programming with no link to TVs
O'REILLY v STEWART
faux news comedy program The Daily Show with Jon Stewart has proven to be a more reliable source than news stations

HQ! IT'S THE COPS

government still doesn't "get" the Internet

Life Cycle

Behind all the jocular criticism of collegiate design programs—or more accurately, criticism of student perceptions of collegiate design courses; or more specifically yet, the willingness of young and potentially entrepreneurial designers to invest in the process and knowledge of a broad-based design education—there are explicit ramifications to the Internet's slaughter of the gatekeepers.

Eliminating the traditional design industry gatekeepers has a dual effect, however, reflected in the rise of journalism programs. Netflix's breakout original programming winner, *House of Cards*, from David Fincher and starring Kevin Spacey, seems to fit with the ongoing trends in quality "TV" gaining eyeballs and quality previously exclusive to film, thus expanding beyond the confines of broadcast networks. Online distribution deals initially negotiated by Netflix with various content owners were at rates that allowed for a quantity of offerings and growth, a curated and designed experience balancing and competing with the "anything goes" world of the pirates. Paired with efficient disc distribution, Netflix did everything Blockbuster could but faster, cheaper, and more conveniently. Blockbuster's monopoly was eagerly torn down by supporters of the innovative Redbox and Netflix. A parallel shift in viewing support for the character-driven, non-FCC-enslaved HBO, and other cable channels, opened the doors for more content diversity. These two factors made *House of Cards* possible, even inevitable. As Netflix's initial contracts expired, the old world business models intervened, prices soared, and the online company was forced to evolve: costs for users increased, popular movies and shows were dropped, and many annoyed customers left. By then, Netflix had permanently changed viewing preferences, helped along by related distributors like Hulu and Amazon, and the most obvious solution was simply more quality content. Viewers clearly supported the storytelling afforded by television as a medium, embraced the shorter seasons and aggressive content of cable shows, and enjoyed the instant access of online distribution. High-level film industry icons, both behind and in front of the camera, were interested as well. The perfect collision eventually led *House of Cards* to the Oscars. It was a complete life cycle; the old guard refused to innovate, were usurped by the masses' support of a few pioneers, and, eventually, some highly engaged members of the previous establishment adapted, creating a broader range of work than had existed before. The old gatekeepers of content, studios and distributors, were still employed and still making money, but they were no longer all-powerful. Hanging over every project was the threat of viewers turning toward pirated means if the offerings did not measure up, not to mention the user-generated content potential of YouTube and Vevo, the opportunity for crowdsourced funding via Kickstarter, as well as the new studio/distributors of Amazon, Hulu, and Netflix. If the old gatekeepers were not dead, they were at least sharing the podium with brash, progressive, risk-taking competitors. There are major intertwined cultural forces at work here, but since they have the same root—the Internet, duh!—we can pause the metaphor.

TV Execs exist in roles similar to their counterparts in design, and their clout is shifting for similar reasons. Design gatekeepers are mostly practicing designers, writers, teachers, critics, or a combination, which is not something you can say about their broadcast-based parallels. This diversity and grounding enables nimbleness and a broader perspective, supporting evolution as opposed to defensive protectionism. Each arena of design has its own threads with requisite gatekeepers. Aspects of the design industry overlap, though some genres have specific avenues to the top: furniture and products often gain more weight from in-person displays, and furniture especially needs a physical space to be shown in, meaning that finding non-online shows can be vital. Education, studio employment, shows and events, design blogs, and magazines all present opportunities for exposure, and all have hierarchies and gatekeepers. Their clout, regardless of genre, has been impacted by contemporary technologies. Part of this has to do with content and message, but mediums and tools are also evolving. Sometimes, the universal desire to be in front of the wave actually leads to curated monotony or fosters work that seems more inevitable than innovative. But giving credit where it's due, many gatekeepers have been more than willing to push for diversity in mediums and processes, as opposed to ostracizing new means and Makers. This is partly why collegiate programs, historically called Graphic Design, have outright gobbled everything tangentially related, or at least built overt bridges to printmaking, 4D, interactive, and installation. Design is an umbrella medium, and Cooper-Hewitt is aware.

No

ENDANGERED
as print publishers like Penguin slashed their offerings, authors embraced outlets like Amazon

FIREWALL
metaphor, gov't has an uneasy relationship with tech companies overseas

A REMIXER'S MANIFESTO

1. CULTURE ALWAYS BUILDS ON THE PAST

2. THE PAST ALWAYS TRIES TO CONTROL THE FUTURE

from Brett Gaylor's documentary RIP

3. OUR FUTURE IS BECOMING LESS FREE

4. TO BUILD FREE SOCIETIES YOU MUST LIMIT THE CONTROL OF THE PAST

Education Gatekeepers

If design education is theoretically supportive of students' need to work across mediums on their own and with others, then a similarly enlightened gatekeeper seems ideal. Historically, design programs pushed theory, history, and craft; but it all flipped toward industry-spec process and form during the expansionist '90s, as the plethora of newly born programs fought over the same scraps. All these programs were patronized by a flood of students as College became What You Did, but the students were ensconced in specialty dorms paid by astronomic loans, and they viewed the whole exchange as a vending machine. For $100,000, opportunity just isn't enough unless it's catered. The more related programs, courses, and faculty, the greater the need to differentiate and parse the details. Entitled students and bloated programs: a blimp competition attended by princesses.

Soon after, the economy bottomed out on a maliciously placed speed bump; arts were cut from high schools, and design programs had to compete for fewer students, many ill-prepared, and all with less money. If the students were entitled before, now the princesses demanded jobs in addition to their suites. As Neil Postman argues in *Technopoly*, reducing education to job training is a culture's dying gasp. The millennial design programs are at least breathing heavily. Dealing with this Lemony Snicket–level of misfortune leaves faculty scrambling to focus more on tech demands and job training expectations, with the recent needs of interdisciplinary and collaborative work; meanwhile, bureaucrats push for an increasingly nonsensical interpretation of accountability. Launching a non-degree-granting, open-access makerspace is hard when the arts are getting cut. A holistic approach is unrealistic when funding is tied to cranking students through in four years, even if they should be aggregating more courses outside their discipline. Faculty are coerced into pushing a four-year plan, even when students decide they are studying in the wrong area, because education is viewed as mere job training and government funding is tied to assembly line implementation. Not surprisingly, administrators treat their schools like a business and refuse to see the benefit of anything that doesn't pair degree acronyms to specific price tags. Interdisciplinary and collaborative design make sense culturally,

INSERT THOUSANDS
get grades → get
diploma → get job

SCHOOL GATE

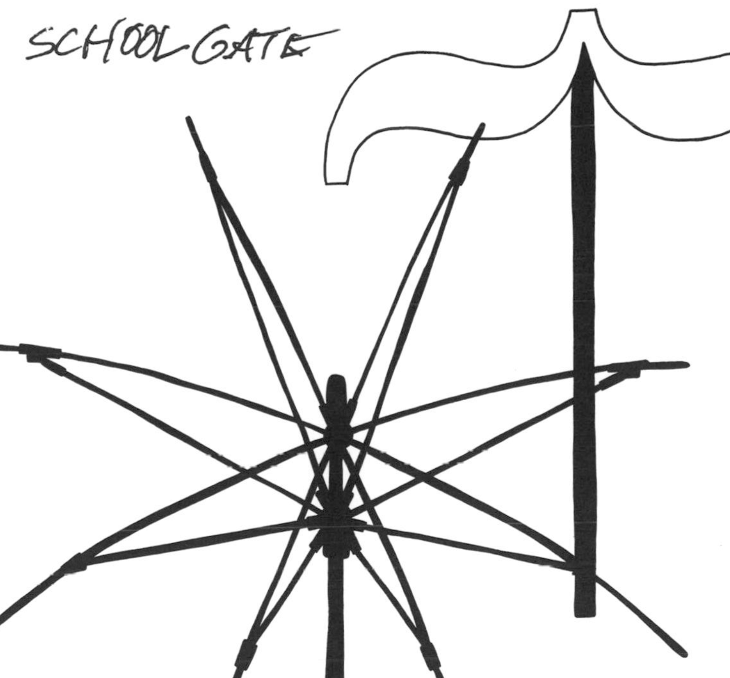

DESIGN IS AN UMBRELLA

a mediumless field now prizes generalists

FACULTY SHOWCASE

increased class sizes, lack of upper-tier positions and job security, decreased funding, and student entitlement increase demands on the entertainers

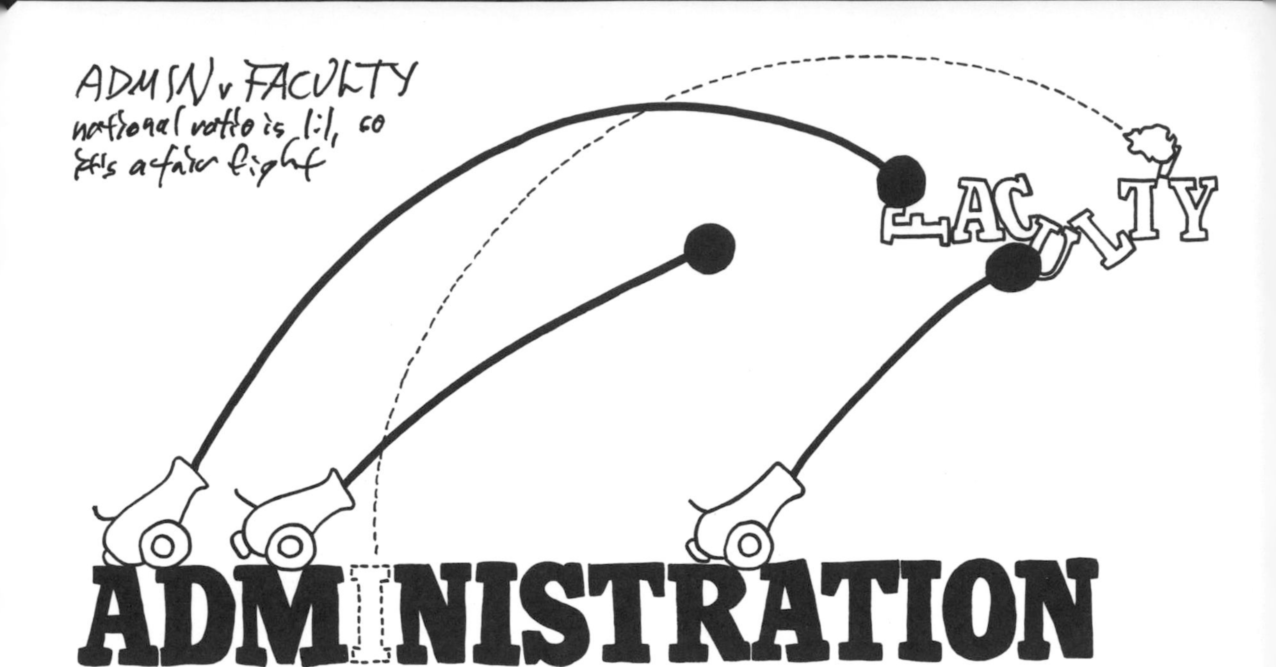
ADMSN v FACULTY
national ratio is 1:1, so it's a fair fight
FACULTY
ADMINISTRATION

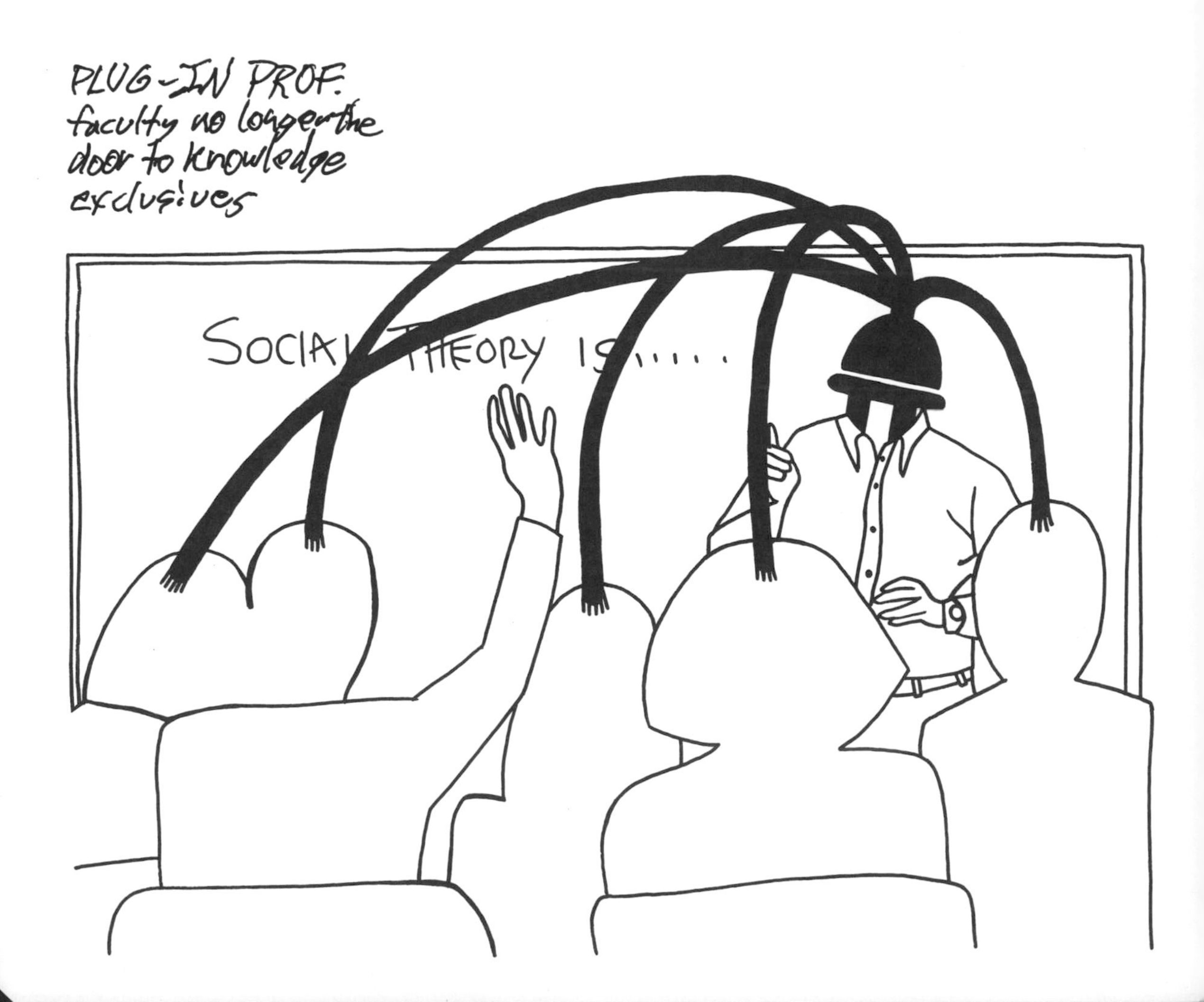
PLUG-IN PROF.
faculty no longer the door to knowledge exclusives
SOCIAL THEORY IS.....

in education, and in practice, but not through a business lens. So while Education Gatekeepers may wish to implement a less top-down approach to art, they fall back on the traditionally defendable.

Structural cracks are partly due to the tenure system. Even though the same system can prevent culturally in-tune turnover by bringing in young blood, it can simultaneously protect faculty who choose to evolve content away from Moneytheism. Design faculty have long prioritized networking, active practice, and importing visiting designers for presentations and workshops. Financial straits make the latter difficult, but Skype provides at least some alternative. In these scenarios, the active rockstar Makers are not functioning as gatekeepers at all, and faculty are primarily table-setters, preparing ingredients and utensils for students to explore through dialoguing. Since art programs have limited faculty and funds in general, the few professors of all ranks do possess some gatekeeper authority.

The overlap of Deity of Education and Studio Overlord is debatable. A high-ranking, well-known, or influential academic is generally considered an authority in their field, but for most disciplines, that translates primarily as publishing credits, either based on research/study or sometimes on the meta-education side of the subject. Because terminal degrees are generally doctorates, intrinsically a research degree, this subverts the potential for crossover academic rockstar success as MFA-Makers. Design programs must develop interschool relationships by educating educators about how design fits into academia, rewriting research policies and definitions, authoring interdisciplinary courses, and sometimes even separating design schools from their art school parents. This can satiate an immediate student body uninterested in diversity of practice, allow design faculty equivalent gatekeeping authority, and craft paths for students and teachers that specifically reflect their views of the industry; however, it flies in the face of the contemporary CO LAB culture.

MONEYTHEISM
when gatekeepers misplace values

All of this raises an impertinent question, one of those What If hand-raisers: "Should education have gatekeepers?" Pink Floyd took it further—*"We don't need no education / We don't need no thought control"*—co-opting a burnbabyburn approach to faculty, readily apparent in modern conservative edu-politics, and it's out of step with today's irreverent interdisciplinary collaboration. ras+e have encountered their share of isolationist intellectuals, but the majority of design faculty seem more focused on promoting connection, and their battles center more on fighting upper administration over the right to do so. Activist faculty are phasing out, and the incomers are more interested in facilitating than leading with upraised sabers. The anti-college movement takes issue with claims that higher education is relevant life preparation, arguing that contemporary access allows people to find their way without jumping through hoops of irrelevance

ALICE COOPER
rockstars persist
but without gatekeeping

put in place by a machine that soaks up students' financial futures (i.e. "unschoolers"). Even when the instruction is formalized through a MOOC or any source of digiknow, the role of gatekeepers runs closer to an informal conversation than obscure knowledge as a rite of passage. Anti-college seems more of a reaction to the '90s inflation in which Everyone Goes To College, than a response to the current universe with its searchable pedestals to Code Kings seated in LCD halls of Graffiti Art. If the system does shrink, and some programs or schools dissipate through outsourcing or neglect, the innovative programs will nominate new gatekeepers for their new structures: evolve and survive.

CHICAGO PUBLIC LIBRARY
reinvented as makerspace

KNOWLEDGE, GO!
wear pajamas to class
even more often

For all the Openness celebrated by Everyone, academia will always have its gatekeepers; otherwise, Harvard would be just another NCAA hoops program. Usually it is tenured faculty at the helm, deciding who gets into the school, into the program, what recommendation letters get written for whom, which rockstars are invited to speak on campus and which students interact with them, what classes are offered with what content, who gets advised to do what or turned onto what opportunity, what connections get made between ____ + ____.

These aforementioned rockstars are gatekeepers in their own right. In an overt, transparent sense, it is their ability to hire designers; but as design becomes increasingly stratified, their status as a Destination for interns is equally relevant. Some of these designers man the parapets of Education and Industry, but that seems to be in decline. Ellen Lupton and Steven Heller are some notable exceptions, but despite an actively large footprint in the cultural design discussions, their publishing records, MFA directorships, and their deliberate willingness to seek means for creating opportunities for others cement their status within Gatekeeper Club. Despite the protestations of the anti-college mob, these aspects of academic/design gatekeeperdom are all positives. Such gatekeeper perspectives as these remain key voices in the collaborative, interdisciplinary, entrepreneurial trends.

Loading

Design is becoming increasingly stratified; and school, with its requisite rockstars, is a major component in that hierarchical chain. The math breaks down as lower barriers to entry for gear (design used to be a tech degree) and knowledge uniformly raise the floor for everyone, which in turn increases the number of designers. Competition goes up, as does collaboration, but the amount of available work can only support so many practitioners. Demand increases for higher-level education to push the ceilings on the crowded field, a demand echoed by the industry and gatekeepers. Lowering entry points eliminates some need for validation, but the ensuing glut just brings the demand right back. In this somewhat decentralized design-hive, the leaders' voices are often amplified partly by education, with its parallel growing block of academics and critics. The need to stand out, to rise on glitter wings rather than the hivemates, facilitates the current escalation in design degrees, further validating the existence of Gatekeeper Lupton and Gatekeeper Heller. Everything gets bigger simultaneously, or else it would collapse under its non-distributed weight like the Street Art Market, though self-taught/made designers and technologists also factor in.

Design Industry and Design Education have always been fairly symbiotic for all the obvious reasons. Standing as a contrasting example, Law School feeds a field noted for the increasing disparity between what is taught and what is needed, creating significant disagreements over the roles of industry and education. Design education gatekeepers are not going anywhere. They possess vital contemporary influence, often coming from the industry with relatively short detours for a two-year MFA. Still, they are not omnipotent or intrinsically necessary as gatekeepers, merely beneficial. Their voice within schools belies the disdain commonly shown toward the arts within institutions, due in part to the increased profile of design within culture and the rising number of student applications. Any potential clout gained manifests less as a Hall Monitor, and more as leverage to build the program: acquiring gear, launching interdisciplinary and collaborative spaces, graduate funding, need-based scholarships, tools for community engagement, and leverage for makerspaces. This would all further facilitate the push for high-floored and high-ceilinged interdisciplinary collaboration, and the launching of small studios, collectives, and start-ups.

instructables.com

THE WORLD'S BIGGEST SHOW & TELL

WIKIPEDIA
even the opposite
of a gatekeeper
can become canon

W

lynda.com

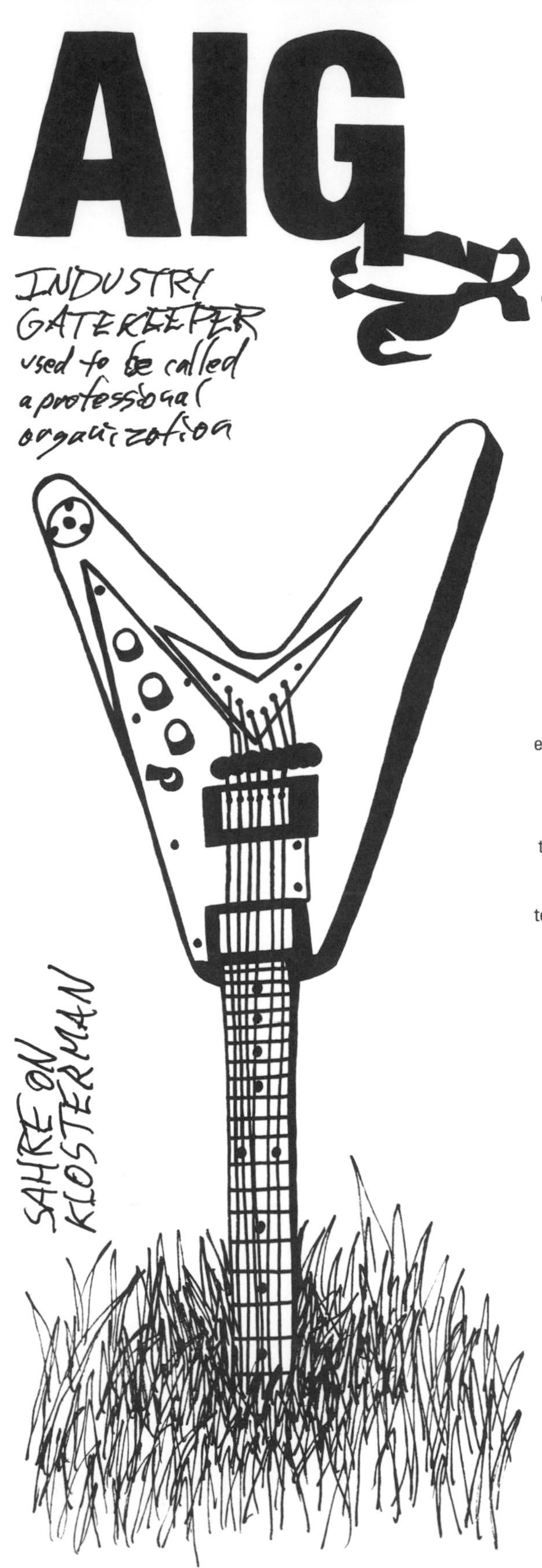

Studio Gatekeepers

Originally these comments were conceived as "Industry Gatekeepers," but the very democracy of tools and information that has broken down the strict hierarchy of the design industry has been complicit in the eliminationblur of Design vs. Art, and part of the overall breakdown of mediums. The small, interdisciplinary collaborative can certainly be discussed as a studio or entity, but it does not intrinsically fit within the semantics of Industry. Even the Fine Arts lens holds here, considering the factory of Takashi Murakami and the prescient pop of Warhol's Factory; a singular vision at the top facilitates the growth of other Makers while deliberately crafting a market for accessible work through a blend of High and Low, Firm or Studio, Invention or Production.

Design's stratification has an overt impact at the industry entry point. The Junior Designer was last seen beneath an asteroid, wingtipped feet protruding from the crater. In a larger firm, especially the full-service dinosaurs, production is an assembly line in constant need of fresh meat, but the democracy of access and quantity of applicants has reduced patience for learning tech on the job, and these lower-level positions are filled by well-educated, experienced designers. In a smaller studio there are few, if any tech-specific jobs. Everyone does everything. The entry bar is high because it can be. Studios now have shifting entry points.

First jobs have become second or third jobs when factoring in the internships that replaced the Junior Designer positions. The cumulative impact is that Pentagram's interns are MFA candidates at top-ten programs. For little or no pay, students become essential designcogs in major studios, working for "experience," saving the firms money, and all to just step onto the ladder—climbing competitively not included. All young designers pursue the same few positions at large firms, as smaller studios are unable to support any form of hiring. These internships are not a step up, they are not additional experience; they are the new normal, and they are essentially mandatory. Stranded on the ladder, chained by 100 grand worth of debt, are the students—as indentured servants to an industry as a whole. Even when the work is engaging and the recommendations stellar, these unpaid jobs are now built into studios'

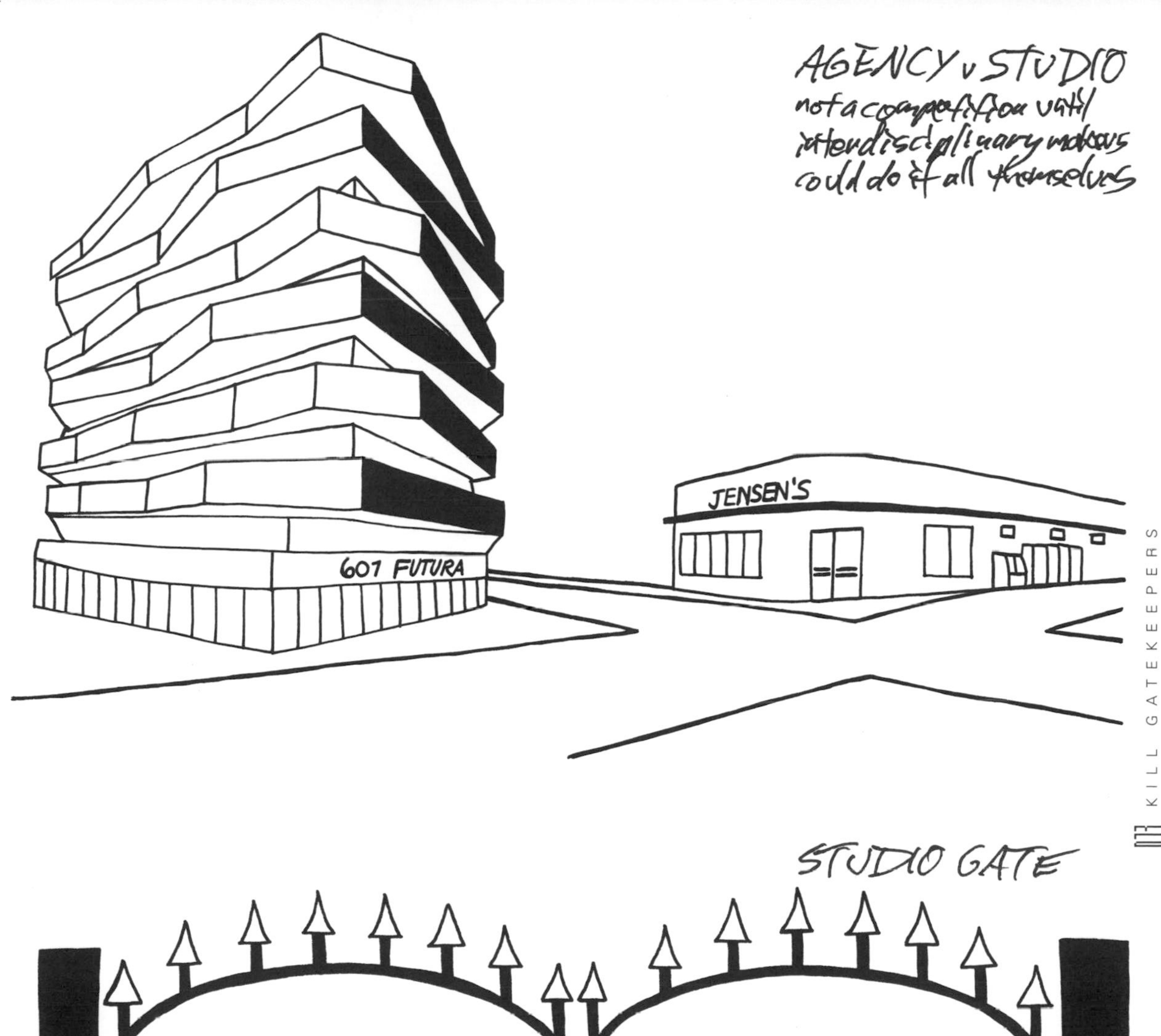
AGENCY v STUDIO
not a competition until
interdisciplinary makers
could do it all themselves
JENSEN'S
607 FUTURA

STUDIO GATE

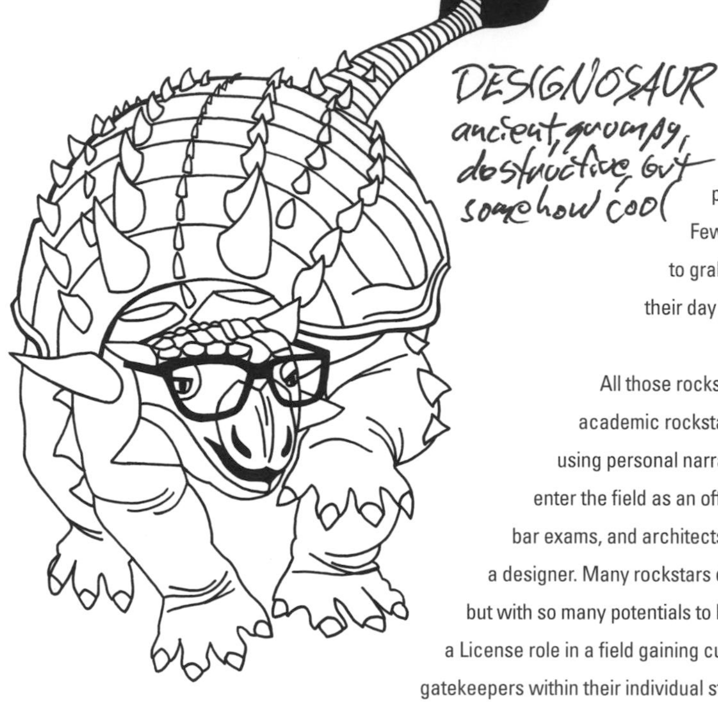

structure and will not become actual positions, especially when the studios are smaller. Internship gatekeepers seem all-powerful and permanent.

But the gatekeepers are dying, fortunately, as practitioners bypass the ladder/industry altogether. Few jobs and much debt prompt many recent grads to grab some friends and start a small studio outside of their day jobs.

All those rockstars who Visit Designer at the dojos of their academic rockstar peers preach a gospel of Portfolio, convincingly using personal narrative to explain that degrees are not necessary to enter the field as an officer, and that the work is all that matters. Law has bar exams, and architects must be licensed, but anyone can call themselves a designer. Many rockstars did gain entry and fame based purely on the work, but with so many potentials to hire, degrees and internships do matter, performing a License role in a field gaining cultural appreciation and clout. So the rockstars are gatekeepers within their individual studios, but hanging a shingle in a part-time capacity has never been easier, and the opportunity to be found a retweet away.

In the classic Silicon Start App sense, college is a perfect incubator for designers to join forces with peers around a singular product or concept, launching it with little to no capital concerns. Designers absorb some of the interdisciplinary bleed through the small-group process and can gain in-house experience while sidestepping the job hunt or internship processes altogether.

Making-based collectives provide similar opportunities, and for those who take the process further to the level of a true collaborative, the voice becomes even more singular. With the collective at their back, designers have the ability to explore a medium with feedback from peers—or build a body of work—leading to a showable, salable output that can grow legs and obtain a following. In this approach, social media often becomes the resume, the point of contact for gatekeepers and others looking for collaborators or employees. In a startup, the ladder can be joined partway up and on the terms of the designer's choosing. Even without working in a collective or collaborative, any precocious 16-year-old with a smartphone, or screenprinting music fan, has an equal opportunity to have their work stolen by Target.

A BOUNCER NAMED WEINGART
let me see your type

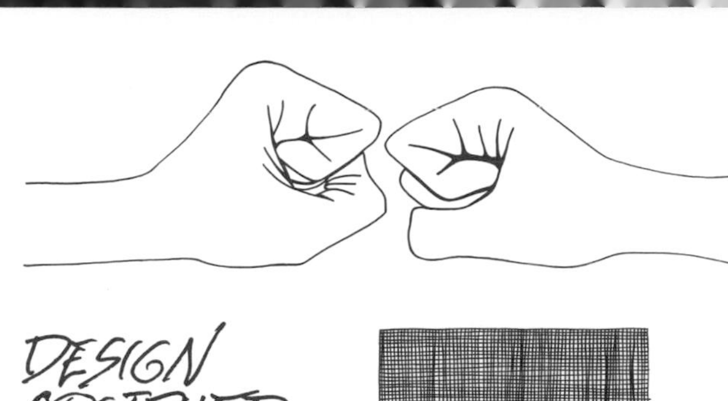

Publicity Gatekeepers

Validation of work and promotion of a studio was, supposedly, somewhat tied to publications and awards. A designer seeking to raise their profile could pay *Graphis* or *Print* to consider a piece for an award or contest or publication or whatever, and then if the magazine deigned to print the piece or give the award, then the designer and their client received a Colbert Bump. This is as gatekeepery as it gets, with the possible exception of Art Directors Club, where membership is fairly literal.

Everyone likes recognition, but it seems fairly outdated even without counting all the magazine foreclosures. Type Directors Club was always a fairly nerdy cadre, the Guardians of the Gorgeous, for whom a pretty and expensive object as output makes sense. But when it's fairly easy to submit writing to *The New York Times* and The Whitney's curators can scour the globe for talent from a couch, the old world status symbols seem less omnipotently hierarchical. Fans of democracy rejoice.

TYPE DIRECTORS
CLUB
guardians of the
gorgeous

When everything is on the table, the curation of learned experts provides a relevant service, even if these folks are no longer perceived to be clad in scarlet caps and seated on high-backs. The gatekeepers still matter because they possess useful perspective, not because they have cultural king-making powers. This becomes immediately apparent when any design student tries to Young Gun their way into a job without interning. Within academia, publishing is still a Gold Standard—which encourages some high-profile examples of scamming by wobbly essay writing and sketchy acceptances—but academics can still translate that seamlessly to post-print. For Makers, the global village seems like an odd place for empirical canonizations of an immense outpouring of subjective, interdisciplinary, collaborative design/artwork. The rapid rebroadcasting of blogs/posts and other online publishing/distribution opportunities provide boosts to designers' Twitter feeds, raising their profile and number of followers, often without the burden of submission fees. Sure, maybe the awards and events were more selective, but that does not make them more accurate. The desire to push further up the ladder—to publish a gorgeous monograph, or write a culturally meaningful essay—ensures that the events, awards, and glossy books will continue. But the publicity gatekeepers are closer to an interactive DJ than a deity.

AN IMP NAMED
WEINGART
were you reading that?

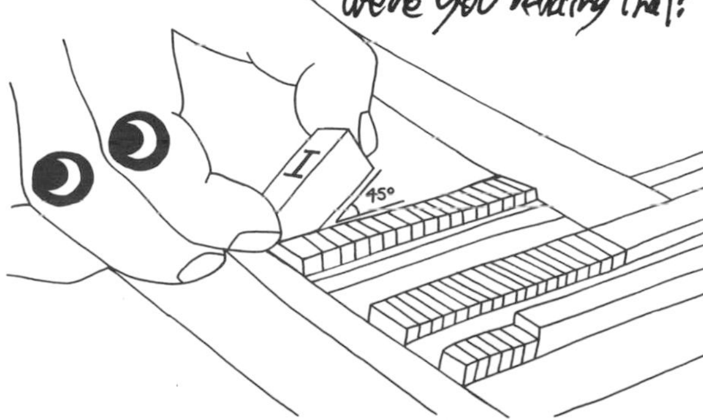

Curatorial Gatekeepers

Collaborative street artist JR summarized the criticisms of the traditional gallery structure during his presentation upon receiving the TED Prize: "The city's the best gallery I could imagine. I would never have to make a book and then present it to a gallery and let them decide if my work was nice enough to show it to people. I would control it directly with the public in the streets." JR's insistence on giving his work to the community without official validation, support, or financing gave him the freedom to evolve his installations to a massive scale that marks his current projects. The politically insistent photography paste-ups are conceptually linked to their locations: the Israeli-Palestinian wall, a *favela* in Rio, and waterproof roof-coverings. Attempts have been made to bring street art into galleries, but for someone like JR, such moves would be conceptual lunacy.

MICHEL GONDRY
tribute to the lo-fi love
of Science of Sleep
and Eternal Sunshine of
the Spotless Mind

Many forces intersect, align, and mate to create an international environment that enables JR's success. The DIY movement includes elements of networking, marketplace, and access to ideas and materials of everything from Web 2.0 on, such as ebay, Fab, and Etsy. This same designwave of interest can be seen in the suddenly high profile of professional NBA and NFL athletes at assorted Fashion Week events, as well as in everything involving David Beckham as David Beckham.

Access in art is cyclical. Exposure to alternative photography, installation, street art, and collaborative work enables a group of young Parisian graffitists to birth JR. The catalyst was a lost, cheap camera he found during an excursion. This access made it possible for him to acquire grants, pipedream money for him when he first started pasting portraits of his friends. It also provided the link to global concerns that previously had been locked behind a TV screen. His work rebroadcasts online, where his photography skills viscerally play a vital role in reinforcing the velocity, becoming an influence on the next wave of disaffected youth exploring their iPhone's camera. Meanwhile, TED hands the artist a prize and pushes his story and project pitch into classrooms and bedrooms.

The gatekeeper role evolves. TED becomes a sort of textbook editor/author/professor/publisher canonizing JR, his ideologies, and his portfolio. It is no Council of Trent, but Luther himself would agree that choosing certain knowledge as IN, not to mention breaking it into blurt-sized bits, can be as powerful as a God-shaped rubber stamp.

A TED invite goes on the CV. A TED vid can be held as Officially Sanctioned Knowledge by the California-style Department of Education. Viewers now know "a something." This easy digestibility of a subject, paired with the respected curation

CURATORIAL GATE
FRANCIS FORD COPPOLA
used fame from early work to finance winery to finance indie films
The Godfather

011 KILL GATEKEEPERS

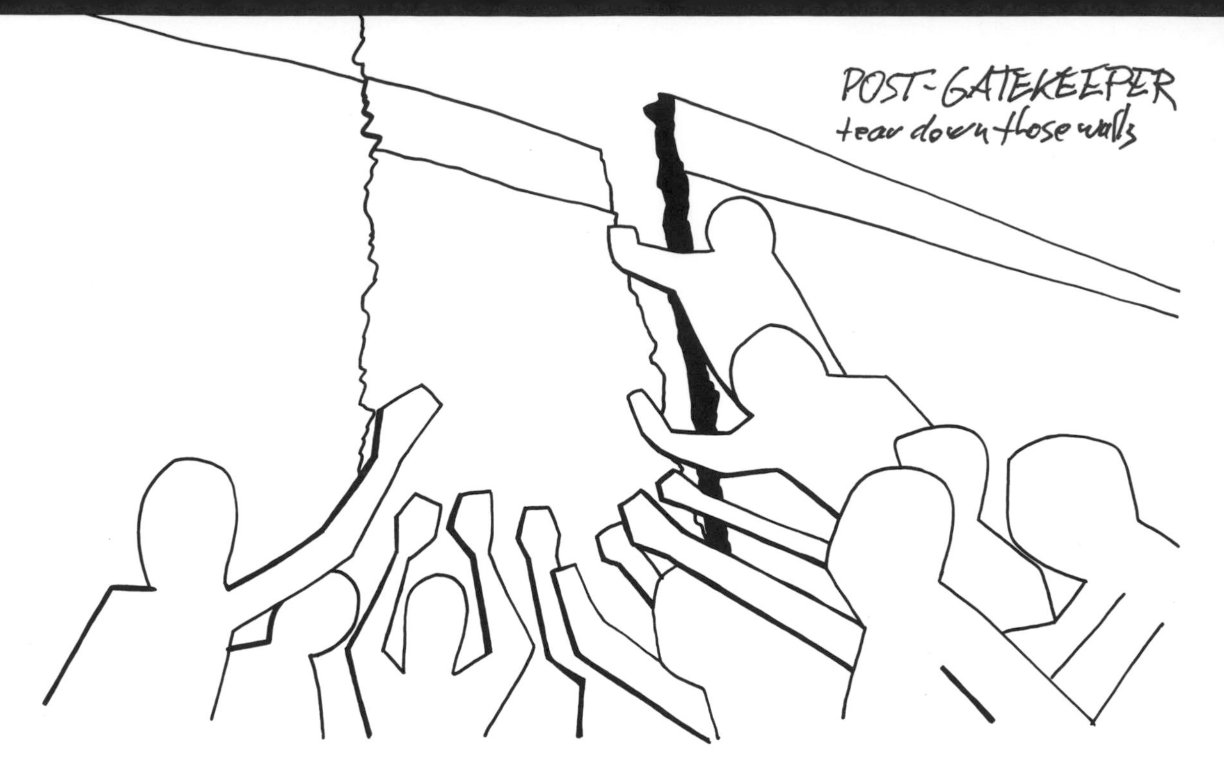

GEORGE CLOONEY
used fame to launch Section 8 with Steven Soderbergh and get non-mainstream projects into production

of a globally endorsed brand, rapidly creates squads of mini-experts. The danger has less to do with any TED-specific flaw, and more to do with encouraging the belief in bite-sized mastery. A counterargument is the related DIY philosophy that these Info Nuggets grow in significance and depth when the audience layers the new information with their own investigations.

DIY Learning enables modern design collaboration beyond the strictures of gatekeepers. The rise in alternative spaces for showing work stems from a rejection of gallery authority, commiserate with democratic means of creation and learning, but it equally has to do with an increased appetite and appreciation for work that does not fit in a white box. As claimed by the many graffiti artists with art school training: "I paint on walls." But paradox, that most truthful mechanism of perceiving reality, rises again in the cycle; this newness is canonized and reinforced by bringing work into the controlled environment of galleries and biennials. This broadening of the post-gatekeeper sphere encourages interdisciplinary collaboration. Groups show work in areas that can be designed conceptually in relation to the piece as a whole, freed from considerations of the "borrowed" room. Work is free to absorb the qualities of the group's manipulated spaces. And the gallery director's concerns of monetized authorship dissipate in the face of collaboration.

We acknowledge the ongoing, if evolved, nature of gatekeepers in design and art, and this variety is an intrinsic positive. Some absolutes do remain. Galleries extend the idea of authorship by publishing artist interviews in their blogs, though the same mechanism can give collaboratives an equal, but joint, voice. *HOW* is no longer the only path to fame. Killing a system that caters to a limited design view will benefit the innovators and collaborators who believe that their voice is strongest under a collective umbrella. Simultaneously, *HOW's* online presence can grow and adapt to this diversity without needing to rely exclusively on the limited viewership of traditional industry vets. In the latter example, the gatekeeper stays, but the curation evolves.

The post-gatekeeper world is faster. More diverse. More collaborative.

The Thing Is...

The accessibility of distribution and authorship now allow flexibility of production, and a bypassing of particular gatekeeper structures—and skipping gatekeeper structures eliminates their much-mocked rubber-stamping. Thus, designers can bypass the vindicating recommendations from faculty. In the flood of DIY design proliferation expanding at the speed of blog, however, the humanization of a body of work via faculty and peers can help provide necessary context within the design industry, as it tends to be an incestuous and self-congratulatory field. Fortunately for those working without the safety net of a degree, design is extremely accepting of new talent, tools, and perspectives, even under the purview of seasoned eyes. Very often, the welcome wagons are driven by gatekeepers excited by the prospect of promoting alternative backgrounds and interdisciplinary experimenters.

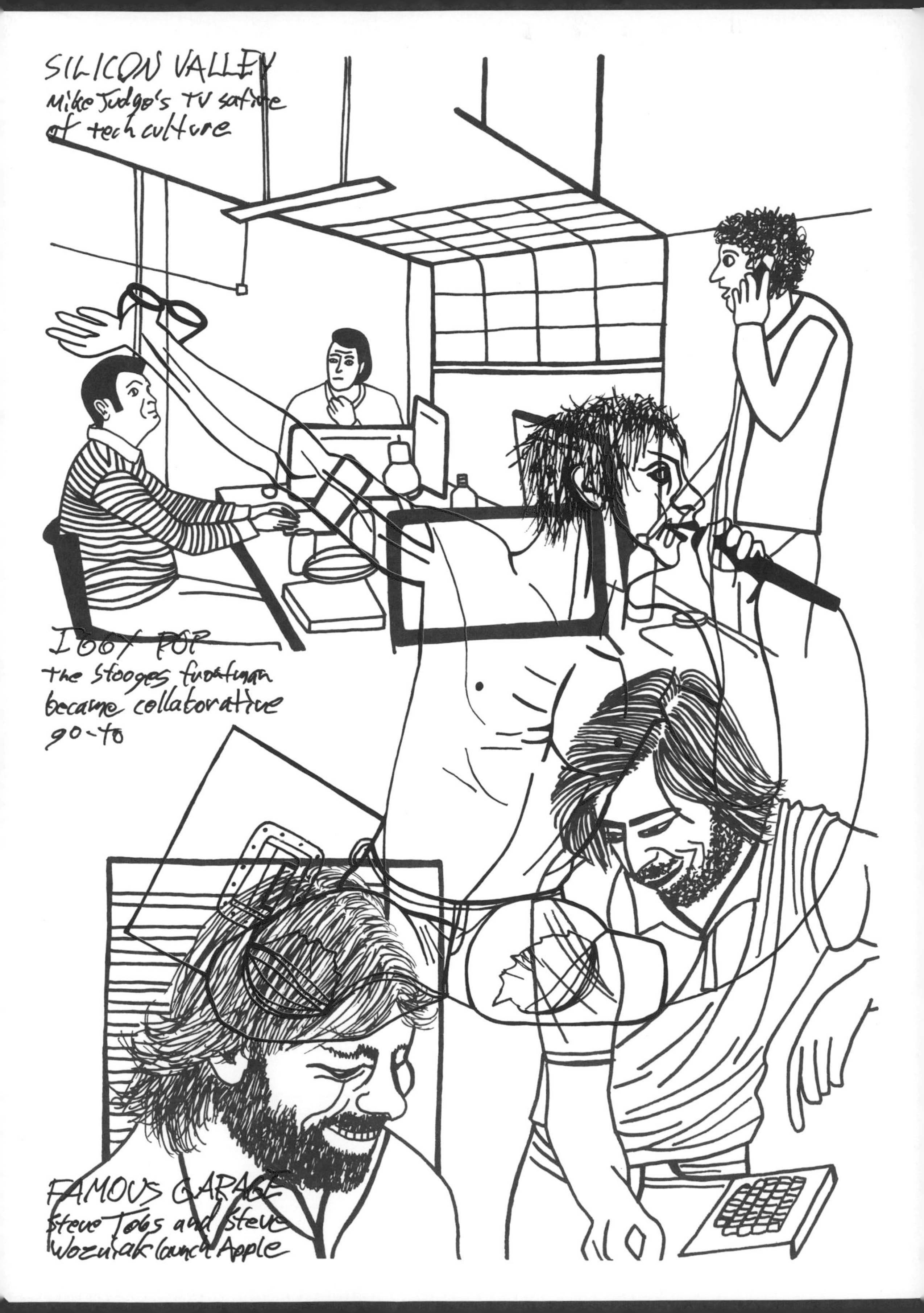
SILICON VALLEY
Mike Judge's TV satire
of tech culture
IGGY POP
the Stooges frontman
became collaborative
go-to
FAMOUS GARAGE
Steve Jobs and Steve
Wozniak launch Apple

Indies- Life on the Fringes

"We're going to learn how to do this by doing it."
-ras+e, in Cover Letter to Proofreader

James Newell Osterberg was raised in a trailer park somewhere in Indian Name, Michigan, making his decision to become a drummer somewhat preordained. But intervention came one vibrant Jim Morrison performance later + Mick Jagger + James Brown + Osterberg's invention of The Stage Dive = Iggy Pop. Music's great Indie Interdisciplinary Collaborator wove a mantle and then wore it permanently.

Iggy is most commonly associated with The Stooges-canonized by The Rock and Roll Hall of Fame in 2009. But even a cursory bio read will give your inner collaborator a bad stage dive's worth of whiplash. Sure, it helps a musician's collab-cred when David Bowie adopts you as an ongoing personal Maker-buddy, superseding even the moral boundaries of rehab facilities. Both legends transcended genres, drugs, and time, embracing their status as statesmen, and moving seamlessly between multiple teams of Creatives. Even so, Iggy stands alone as a Collaborator of Note due to the breadth of his interdisciplinary work, seemingly bringing out the very best of each inventor he works with.

He sang on "Punkrocker," the Swedish electronic group Teddybears' best song, which feels like custom Iggy. Bowie's "China Girl" was actually an Iggy Pop song. In fact, Bowie's recordings of Iggy songs helped the latter out financially with royalties during some particularly tight times. The Sex Pistols' Steve Jones worked with Iggy on the cult film *Repo Man*. Iggy sang with Lou Reed in the animated film *Rock & Rule*. "Candy," from Iggy's album *Brick by Brick*, was a duet with Kate Pierson of the B-52's and his most commercially successful song to that point. Green Day, The Trolls, Peaches, and Sum 41 all collaborated with a reunited Stooges on *Skull Ring*. Madonna requested The Stooges to perform her songs in her place for her Rock and Roll Hall of Fame induction concert. Iggy sang on the Danger Mouse/Sparklehorse track, "Pain." Slash's solo effort featured Iggy on "We're All Gonna Die," and Ke$ha tapped Iggy for "Dirty Love," proving the icon's penchant for making everyone around him better, regardless of genre. Some additional collaborations include several early Johnny Depp film projects, voice work including Lil' Rummy on Comedy Central's *Lil' Bush*, and a give-and-take role with Marjane Satrapi *(Persepolis)*.

Iggy Pop initially gained attention through unexpected musical innovations, no doubt helped along by performances that included rolling around in broken glass, but his ongoing footprint has much more to do with finding conceptually companionable projects with colleagues who extend beyond his prototypical dominion within American Punk. In cases of projects with Satrapi and Bowie, all parties take turns initiating work and inviting the other as a contributor, as opposed to a hierarchical or medium-specific organization of Creative and Labor. Given that so many of Iggy's projects, musical and otherwise, occurred outside of a major label or mainline pop trend solidifies his status as an Indie outsider, but his ability to make everything he touches his collaborators' best work makes him an exemplar collaborative Indie link.

Goodbye Gatekeepers: Genres Are Gone

The truth is, music hasn't had a real use for the term "Indie" recently. Whether it's a technical definition, the lack of a "mainstream" (major) label, or a broader understanding of an inherently outsider genre, the Internet has brought in new business models that have smoothed out the idea of High and Low art by any traditional music definition. The outsiderness of something based on perceptions of intrinsic medium quality—battles long waged in other disciplines, including comic book illustrators like Daniel Clowes, Dave McKean, and Scott McCloud, as well as fine/commercial artist Takashi Murakami—has moved to the background as everything is now laid out on the remixer's table. Availability and access cut down walls of the taboo, allowing Irony Culture to pillage the distinctiveness of everything. For music, and especially for naturally hybrid genres like rock, the mainstream label benefit has diminished while production and dissemination gear proliferated, but the overall pool of money has vanished. Bands survive primarily by touring, not recording. All of these threads have opened the doors to widespread innovation since labels no longer exert control brought on by Protecting the Investment. In short, invention has been pushed by Indie music, but now many bands flit between genres, have equal access to the Great Popularity Pageant, and create without financial decisions impinging on creative ones. This entire arc of Indie music has parallels to Indie graphic design.

PUNK'S FLAGMEN

The contemporary music landscape features many highlightable sea changes and innovations, but what was once the Indie domain of pushing envelopes everywhere now has little to do with labels or even music genres. OK Go achieved a success-bump when a lo-fi video of the prepped-out band performing a choreographed treadmill routine set to their song went viral, launching them as video artists as much as musicians and leading eventually to a gig at the Guggenheim. For perspective, Baltimore's Animal Collective also performed at the Guggenheim, and they epitomize the Elder Statesmen of Indie Darlings. OK Go does represent that the new Indie music innovations are often interdisciplinary and that their outsiderness transcends genre titles.

tribute to the famous music mag's cover by John Holstrom

Hip-hop group Death Grips' great accomplishment, according to *Spin,* is actually deleting themselves from the Internet. Jack White set up a self-serve recording booth in a local Tennessee mall. EDM-pioneering Daft Punk ditched their now mainstreamed digital sampling for live performances with both Pharrell Williams and Stevie Wonder at the 56th Grammy Awards, promoting an album that garnered attention primarily for stunning inter-genre technical recording prowess. Girl Talk's hyper-sampled compositions, notwithstanding access to all media and cheap tools, have made remixing, both literally and figuratively, the new normal.

Innovative design collaborators also push design by working outside of traditional medium and genre definitions, making the concept of specialization less relevant than exploring interdisciplinary systems, considering 4D and interactive as layers instead of ends, and a general affinity for sacred-less remixing. The explosion of graphic design's popularity, like journalism, is partly due to the availability of digital tools and the interest in exploring how to communicate with those tools. Designers have dropped the need for extreme legibility, uniformity, and printability in relation to expressiveness, branding, voice, and usability; in short, craft vs. experimentation. This influential outsider-voice mirrors Indie and Punk music as a grassroots creation. Meaning, smaller studios and collaboratives are now exerting as much influence on the design industry as mainstream big boys like Pentagram and JWT. Just as Lorde can blow up in a month, so too can Infantree. Without gatekeepers, impacts beyond the Indie's modest financial footprints gain traction even within the mainstream: Andrew Blauvelt's taping mechanics for the Walker's type-able kit of parts identity shares design-DNA with 2X4's Brooklyn Museum of Art blue bubble B system and Experimental Jetset's Responsive W for the Whitney. These anti-classical monolith approaches to identity development bring this Indie vibe to large, mainstream cultural institutions.

Design, like music, may be led from the fringes, as the lines between Insider and Outsider vanish in light of medium accessibility. And because everyone in the design industry has access to the same equipment: the kinds of experimentation, mutual savvy, and informal conversations that are happening in smaller studio collaboratives with curious, angryoungandpoor designers is proving unique, trendsetting, and searchable.

The issue with the Indie label is that the lines between Outsider and Insider are thin and perhaps irrelevant. In music, inter-genre work is happening in an environment where the impact of major labels has dissipated post-Napster. In design, the trends toward interdisciplinary and accessibility trends have had a similar impact: Post Typography can compete from a Baltimore studio with a collaborative that is half design and half illustration by training.

They say I'm just a stupid kid
Another crazy radical
Rock 'n' roll is dead
I probably should have stayed in school

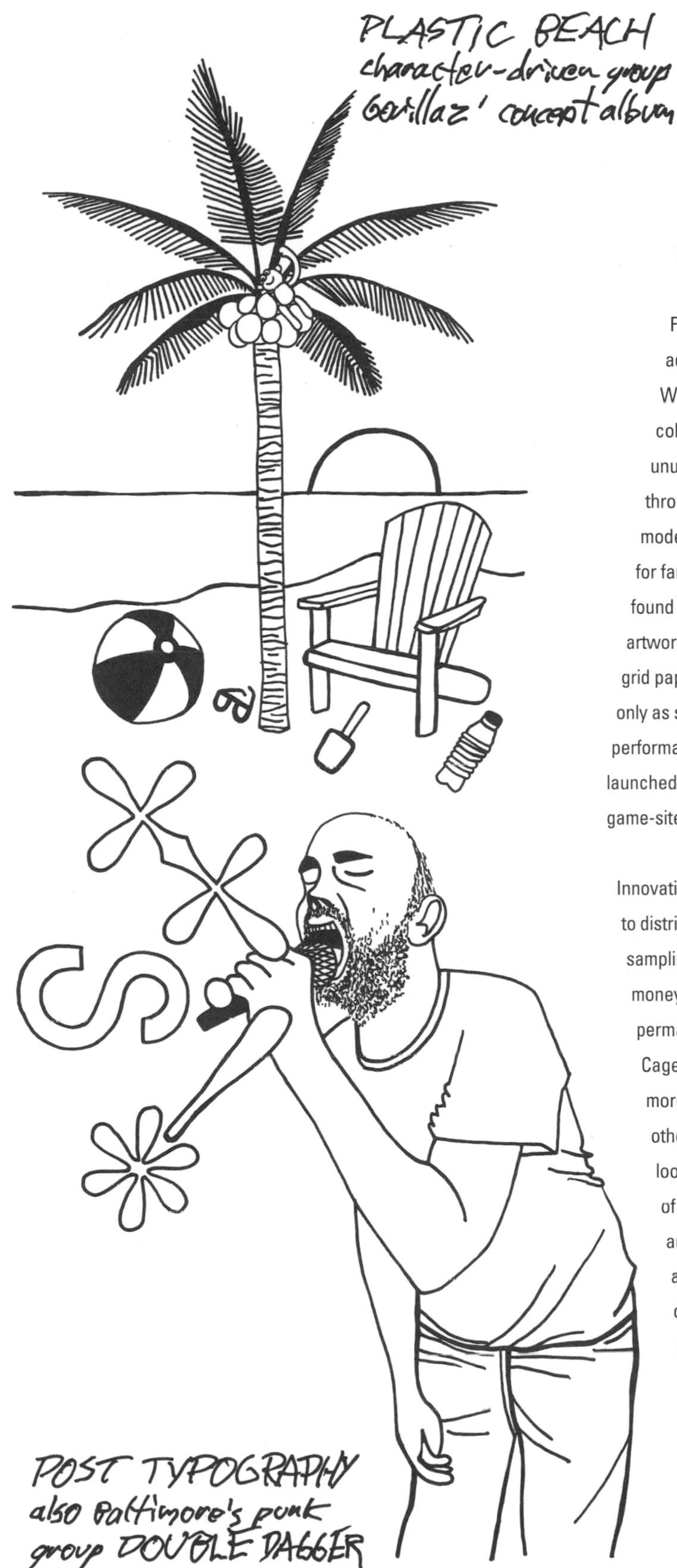

sang Cage the Elephant on "In One Ear," a cheerful "screw you" to anyone who felt that Industry and Numbers drive Content or Innovation in the contemporary music scene. This is the Post Typography attitude regarding the Death of Print as they drop Vimeo process videos of their letterpress posters.

Furthermore, collaborative Indie design takes full advantage of the low post-print startup costs. Without the availability of money and rules in a collapsed label-scape, Radiohead continued to find unusual ways to put their music in the hands of fans through the first sanctioned Pay What You Want release model of *In Rainbows* and by uploading raw bits of footage for fans to make their own Radiohead music videos. Beck found similarly Indie methods of musical reach, with album artwork for *The Information* that was a big, folded sheet of grid paper and a raft of stickers, as well as an album released only as sheet music with a site for fans to upload their own performances that became the Official Recording. The Gorillaz launched a satirical "mind-numbing, mission driven" interactive game-site to sync with their album *Plastic Beach.*

Innovation has not been tied exclusively, or even primarily, to distribution costs and means, but also to remixing, sampling, and repurposing. The openness, and lack of money to drive decisions, has meant that rock 'n' roll is permanently alive, both for throwback practitioners like Cage the Elephant and remixers like Girl Talk. Rock is more diverse and more vibrantly progressive than any other contemporary music genre, though the form is so loose that this allegation is cheating. The malleability of rock, its history of collaborative interaction with the audience, and its huge range of performance practices and historical musical influences all contribute to its diversity. This creates a wide-open landscape for experimentation and play, so long as the goal for the Makers is not "Be Rich." Much of modern music, and design, can be described as Indie, both in spirit and actuality. It's just poor. The Indie designers reflect this. The Heads of State gains traction as contemporary printmakers, though the financial advantage to such a career is zilch.

Go Welsh promotes community interaction by installing donated pianos, painted hot pink, on Lancaster, Pennsylvania street corners. And printmaking is the hottest field in art schools, as frustrated designers finally give-in to their repressed inner Makers. What the Internet does not provide in interdisciplinary collaborative opportunities, makerspaces and hackerspaces provide; hungry designers gain the room and tools they need to share an Indie collaborative community outside of corporate support.

But style, influence, and accessibility pale in the Indie design quiver compared to the Gold Standard of design stars, the bazooka of cred: Bigass Clients. As *Mad Men* makes abundantly clear, the success of designers is defined more by the financial success of their clients, not necessarily the work. Success is judged by the outright Scale of the company. In the AMC show, the ad firm, Sterling Cooper & Partners, breaks every moral boundary imaginable in pursuit of the big boys: car and airline clients. The logic is that having Coke on the resume is more important than the actual design work for Coke. Historically, the (design) public only associated specific work with specific creators once the creators were famous enough to warrant a tome with THEIR name embossed on the cover, so piggybacking off the reputation of a client was the indicator of success for the non-names. Paul Sahre's posters in Baltimore, an attempt to innovate and stay sane while working on mundane client work, catapulted him to New York and the School of Visual Arts (SVA), becoming an Indie Darling in the process. These are rare occurrences; big studios usually gobbled up the young Creatives like Google and the NSA does today.

Goodbye Groupie: Popstars vs. Indie Darlings

The genre-bending and genre-iterating performances of Iggy Pop made him a weirdo Outsider Icon, whereas David Bowie's similarly bendy innovations pedestalled him as pop's Fearless Leader, so much so that his music was used as a touchstone symbol in *A Knight's Tale,* parallel to Queen's "We Will Rock You." Design has its own innovative Indie rockstars, whose footprints sometimes become pedestals as they ascend thrones as Popstars. David Carson's influence was massive, sure, but it's now distinctly dated, pinning the artist as a New Wave genre figurehead, more than a continuing contemporary innovator.

The same goes for Popstar James Victore and his signed posters, snowed under by the enormous quantity, diversity, and experimentation of the hybrid illustrator/printmaker/designer-as-artist-types producing voice-heavy band posters with no hope of much financial or career-boosting windfall. Music's Indie Darlings got swallowed by iTunes and New York, whereas design's Indie Darlings have usurped the thrones of their elders. These small interdisciplinary collaboratives are devoid of fame-seeking Single Shtickness and are instead empowered by an internal push-pull dialogue, consistently articulating new post-box horizons. Sagmeister & Walsh is a great example of a constantly innovating and evolving Indie collaborative. In fact, Stefan Sagmeister takes sabbaticals from his studio work to further his research, like all good Indie rockers writing new work. Meanwhile, many of the early pioneers perch grumpily, stencil-branding their own soapboxes. The name-branded '90s typographers Mr. Keedy, Ed Fella, Zuzana Licko, P. Scott Makela, and pop-activists Shepard Fairey and Jonathan Barnbrook, rockstars who became Popstars, are now accompanied by nimble-fingered collaboratives known less for pure genre mastery and more for genre collision brought on by a collaborative view of authorship. Rob Giampietro is a Literature and Graphic Design alum of Yale University whose collaborators at Project Projects do things like create and tag typefaces for glow-in-the-dark exhibition signage at a vacant meat packaging warehouse.

If graphic design has its rockstars, surely it has Indie Darlings: Iggie Pop rolling through broken glass is Stefan Sagmeister's X-acto-lettered torso. The fact that Sagmeister, Martin Venezky, Marian Bantjes, and Candy Chang have established successful design careers despite work that bucks the GetMeACommercialJobAsIDefinedInMy InternetBrainLust-addled thoughtprocesstrends of many modern design students is somewhat astounding, and it does reinforce the idea that producing quality innovation has no intrinsic tie to financial compensation or clientele. The explosion of interest in design has many threads, but part of the rising numbers of graduate students, small studios, and non-commercial projects, is driven by a belief that unexpected and self-guided Indie work is

valuable, even if there is nary a CEO in sight. That designers of all vibes are able to get their personal projects into the public mind through *How* and other curated blogs is obvious. But the plaque of Designer as Popstar With the Book Deal has given ground to any good hook getting a voice via self-publishing. Even Amazon has become a legitimate opportunity for innovating authors who have been rejected by traditional and established publishing companies. The Work is now the primary standard of Indie cred, thanks partly to tweets and posts, as opposed to an inflexible and trademarked Standard of Arrival.

ANGRY YOUNG AND POOR

Yet as modern communication democracy raised the design floor and lowered the design ceiling, design's accessibility manifested as demo tape havens, like DeviantArt, building a sense of community, collaboration, and Copyleft perspective among designers at an early stage. By the time they hit design school, where the winnowing begins, many ideals are already in play. Since everyone has the ability to make, promote, distribute, and access projects, the Indie alternative to major studios has led to a wealth of small collaborative studios and startups. This Indie-cred proliferation helps combat stratification manifested by the sheer number of newly minted MFA designers interning at JWT, desperately hoping to climb a traditional ladder. While the mainstream Majors will continue to find success, the Indie designers, photographers, and writers are banding together in basements, reaching clients and building highly tailored relationships, all without shipping up to NYC or LA. As an Indie, these small collaboratives eschew the pursuit of shiny clients in favor of particular jobs that allow them to do the kind of work they're interested in. Businesses seek out these highly specific alternative voices. In this landscape, Coke is only a gold standard if they want the ads typeset in a font made from dehydrated syrup, custom built and installed by Chuck Taylored Indiedolls in an abandoned warehouse via flashlight. Traditional agencies cannot afford to touch such projects, but for an Indie with boots planted amongst the masses, it's a dream job.

Because of the McMEGA Client as Design Standard, BFA programs have been slow to reflect the Indie Design of Now, but some are shifting. Schools still advertise their faculty based on a list of clients,

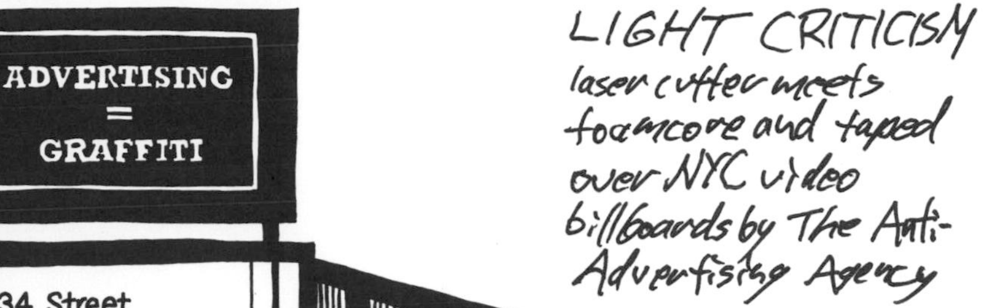

LIGHT CRITICISM
laser cutter meets foamcore and taped over NYC video billboards by The Anti-Advertising Agency

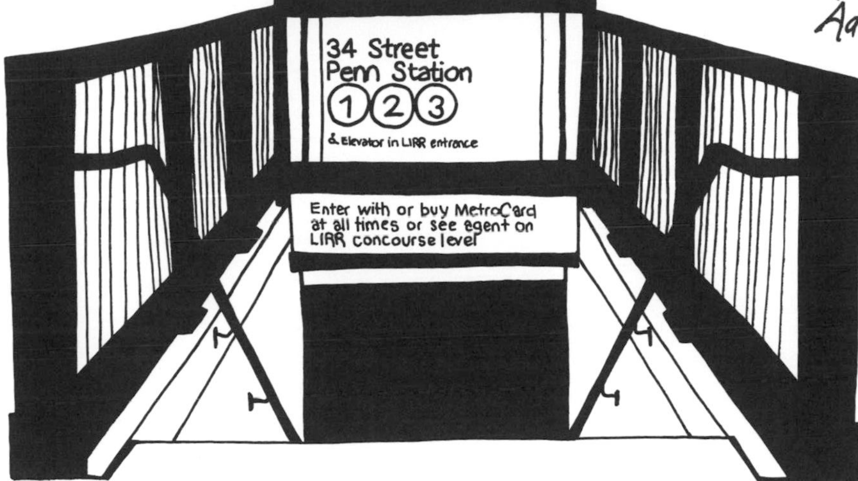

SWEDING
michel Gondry's film tribute to lo-fi art and community launched a genre

YEAH YEAH YEAHS
FEVER TO TELL
tribute to the Fever to Tell album
Spike Jonze and Karen O and Dave Eggers tackle the Maurice Sendak book
ROLLOVER FACTOR
tribute to Howe Industries

as if that indicates anything at all about the program, courses, professor, or their work. It's the equivalent of listing the number of papers an art history professor has written without mentioning what they are about. Listing clients does not even convey medium information about a designer, whereas art history faculty usually have their lines of inquiry advertised. But as schools launch open-ended spaces—for collaboration between designers, for launching startups, for interdisciplinary work that often extends beyond assignments—the value of Indie design may be starting to shift within academia. Perhaps the still unconquered bastion of industry control is the prevalent internship model, which downplays the role of Indie collaboration and voice-heavy, open-ended production in favor of slaving as a support posse for Pop Stars. Incidentally, the resulting inequality of job descriptions creates poor soil for growing as collaborators. Sans investment, agencies are often uninterested in the interns as full-on members, acting as if the internship is a favor for the youngsters. Realizing this trend, many smaller studios have made an overt point to adopt interns as team members. By treating school as an opportunity for fostering startups and student-run studios, faculty are attempting to more accurately and holistically reflect real world collaborative experiences.

Anything You Can Do, We Can Do Without You

Director Spike Jonze asked Karen O of the Yeah Yeah Yeahs to do the soundtrack for the 2009 film *Where the Wild Things Are.* He also asked Dave Eggers to do the writing. Maurice Sendak, the original writer and illustrator of the kids' book, served as a producer on the film. Beyond the fantastically engineered costumes, custom-crafted music, and handwritten credits, what makes this Indie film unlike any other blockbuster arose out of Spike's commitment to the piece above studio pressure and commercial endorsement. Disagreements with Universal prompted Spike to take his film to Warner Brothers to finish. The result was one of the most important and poignant children's films in history.

Jack White apprenticed with an upholsterer for three years, writing poetry on the inside of fabric and practicing guitar in his first band, The Upholsterers. He slept on the floor of his Detroit apartment without a bed to make room for his drum kit. With his first major break, The White Stripes' self-titled album, Jack forced the small Detroit-based label, *Italy Records,* to alter their green sticker, claiming that if they didn't understand the band's peppermint color scheme, then they had no business working together. Jack White now manages Third Man Records, collaborating with many musicians on solo projects, The Dead Weather, and The Raconteurs.

The point is, sustainable collaboratives are not at the mercy of anything or anyone anymore. When rock 'n' roll terminated in its logical extreme with KISS, musicians were able to regroup and reinvent the industry. This is exactly where graphic design is at thirty years later, and in the fringes we're beginning to see frustrated, burnt designers coming together, armed with cardboard and assorted caps, tweeting their hearts on their wrists. Collaborating with one voice may mean not collaborating with another. Indie musicians established the benefits to money-less priorities as a way to carve out a distinct platform, and design collaboratives are following them onto the stage.

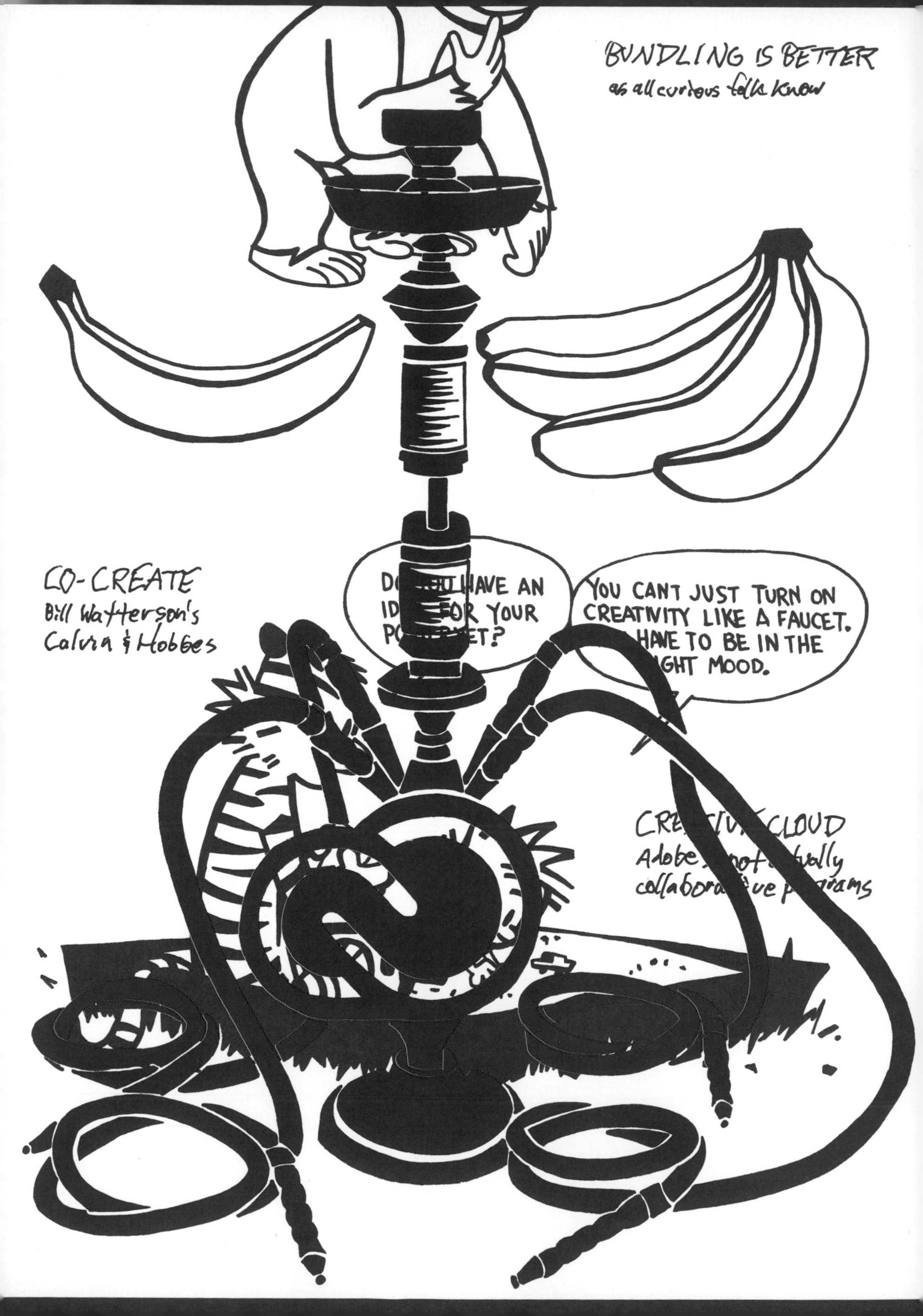
BUNDLING IS BETTER
as all curious folks know
CO-CREATE
Bill Watterson's
Calvin & Hobbes
HAVE AN FOR YOUR
YOU CANT JUST TURN ON CREATIVITY LIKE A FAUCET. HAVE TO BE IN THE GHT MOOD.
CLOUD
Adobe
collabora

"I think we dream so we don't have to be apart so long. If we're in each other's dreams, we can play together all night."-*Calvin & Hobbes* by Bill Watterson

We will be forcibly removed from the big container[1] of infinite fun and loaded into Mom's pressure cooker of a Caravan one hour from now. A sloshing clump of sand at the bottom of our sneakers will be our only souvenir, and to our dismay, Mom will make us bang the canvas urns upside-down against the spare tire before departure. Our socks will be used to dislodge all remaining granules stuck 'twixt toes. Eventually, we acquiesce, quickly force-quitting the burning macadam that turned our pink flesh to yellow callus in a matter of seconds. Worse than wrecked skin, in a few hours, the mountain will be gone.

Our mountains are truly impressive, and I'm not just saying that because I helped make them. They grow from nothingness[4] to dominate the bare terrain.[10] The other schoolchildren build wussy mountains-their speed bumps a reflection of low level culture. Overly ornate speed bumps. To clarify, there are two schools of thought on this: sandcastles should either be lavishly detailed and oozing Gothic icing, or they should be sensationally imposing, "like a giant dildo crushing the sun" to borrow from Beck's "Pay No Mind." We make the latter: huge, loose piles that give the astronauts something to photograph. Other kids' sandcastles are an embarrassment to childhood. In their defense, there are no earth-moving apps yet, so dirty hands are an imperative, which they find confusing.[9] Plus, finishlineless projects perplex the linear-thinking kids.

Danny and I[2] mix in clay and water from the fountain[8] to make a wet slurry, the clay being our secret ingredient. The sticky wetness prevents the mountain's weight from collapsing in on itself when we plow a tunnel straight through its guts. Adding mud is technically against the rules (there's even a sign that says so) but those external forces aren't part of our world.[3] Playground rules are either internal or irrelevant, indicated by the middle school skaters lurking nearby. Running back and forth to the water fountain is probably the hardest chore, but I usually volunteer because the heat raises it to the stature of a water park.[6]

The work goes well and we work effectively, though for some reasonless reason Danny perpetually stabs some kind of wimpy, useless twig into the top of Sandy Olympus. Danny's sense of aesthetics has yet to resolve, but he's a cool kid, so I deal[5] and develop the mountain's ceremonial backstory involving an anemic race called the Twigger Folk.[7] Danny and I build the mountain, then Reuben and his gang take over after we leave, turning the sand patch into a model of the Marianas Trench. This usually means we have to start from scratch when we return, and we grumble out of principle, but secretly we're glad for the excuse to make something even bigger and louder the next day.

(1) Forced encounters between disciplines and practitioners; (2) Physical proximity of colleagues; (3) Open-ended production outcomes and deadlines; (4) Environmental liberties through spacial and structural malleability; (5) Intellectual diversity; (6) Play element; (7) Distraction or excuse element; (8) Accessibility of gear, tools, supplies; (9) Safety zone and freedom to fail; (10) Inspiration and motivation

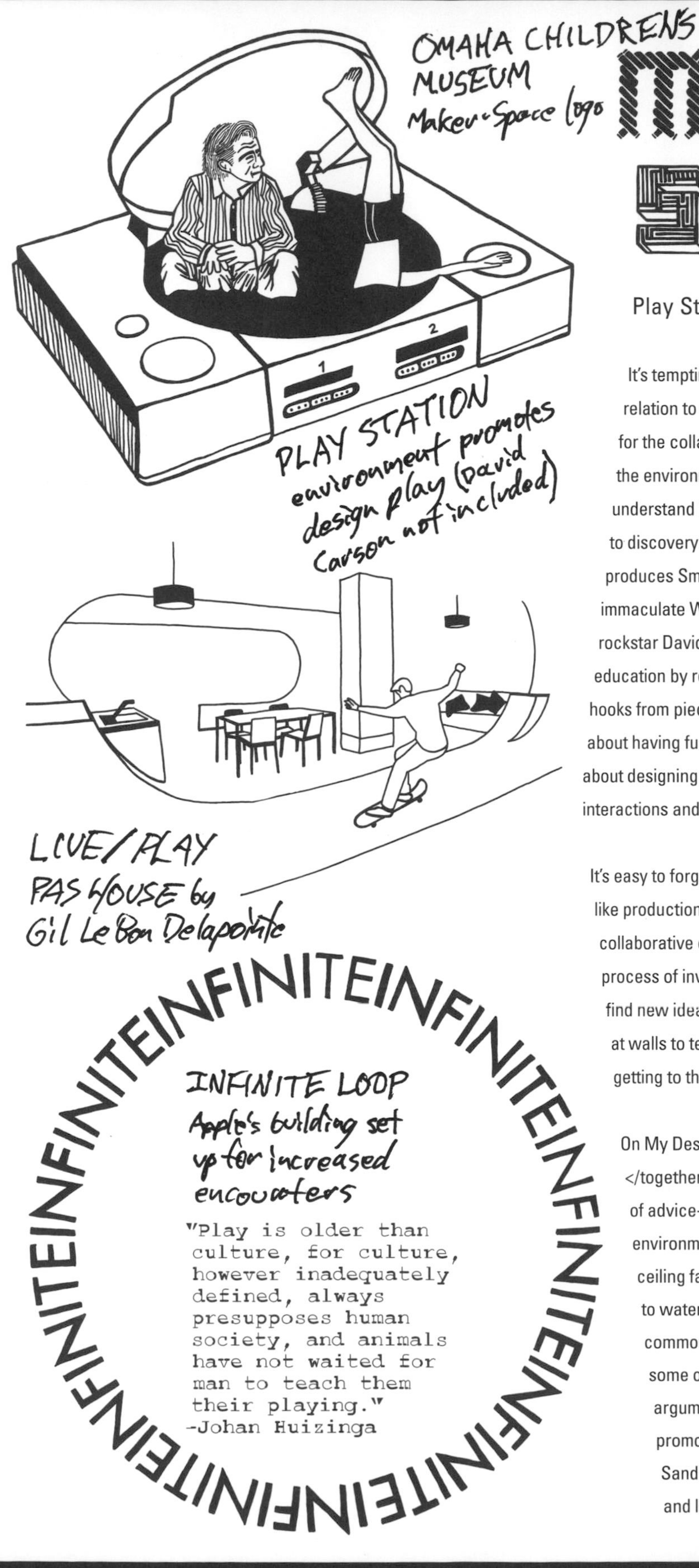

Play Stations

It's tempting to consider Play as a concept in relation to design environments as purely the search for the collaborative cure-all, Fun. As if, dissecting the environments of design collaboratives will help us understand how the walls influence the people, leading to discovery of a mythical kernel where the Playground produces Smiles, elegant Interactions, thus birthing immaculate Works. Sometimes games are employed: rockstar David Carson flaunted his lack of a formal design education by refusing to recycle formal design jabs and hooks from piece to piece. But Play isn't intrinsically about having fun, even though play often is fun. Play is about designing environments that foster collaborative interactions and idea-sparks, enabling fringe thinking.

It's easy to forget how to play. Designers who feel like production cogs may forget the rush of a highly collaborative environment, enjoying company and the process of invention. Playing is how designers work to find new ideas through trial and error. Throwing things at walls to test stickiness carries the illicit thrill of getting to throw things at walls.

On My Design Wish List of life-simplifiers: <working> </together>. Unfortunately, there is no magic list of advice-bullets for increasing collaboration via environment: put tables on 45° angles, install large ceiling fans in a modular grid, install sprinklers to water designers biweekly; but looking for commonalities of effective facilities can distill some overlapping traits. While building an argument for the role of environments in promoting interdisciplinary collaboration, Sandboxing parses example groupings and lists the threads after each section.

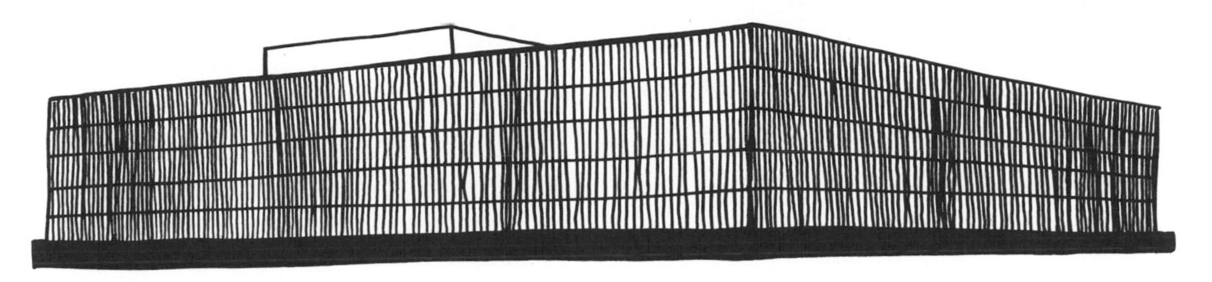

BELL LABS
idea incubator by
Eero Saarinen

FORCED CONNECTIONS
Allied Chemical wanted
interactions

"Some of the hallways in the building were designed to be so long that to look down their length was to see the end disappear at a vanishing point."-Jon Gertner

Bell Labs + Infinite Loop + Yahoo! + Eyebeam

Whiz kid Mervin Kelly, from the mining town of Gallatin, Missouri, missed the technological buzz of the Industrial Revolution, not to mention the hyper-accessible, hyper-connected, hyper-collaborative Information Age away in the future. Over time, Kelly absorbed new ideas regarding the tech magic of radioactivity and X-rays, worshiping the innovation cultures of Henry Ford and Nikola Tesla. Culture was evolving, highlighted by the World Fairs, which began shifting emphasis away from novel technologies to a showcase of socially conscious ideas.

Fast Company's editor, Jon Gertner, writes in *The Idea Factory: Bell Labs and the Great Age of American Innovation* that, while his peers invested in mechanical skill, Kelly gravitated towards the *intellectual,* a person who was able to explain how and why the machine worked. Both influenced and influencer of technological advances, Kelly became a master at fostering innovation through collaboration during his chairmanship of Bell Labs from 1925 to 1959. Under Kelly, patents were received for the first transistor (the building block of digital communication), silicon solar cell, laser light, communications satellite, cellular telephone, fiber optic cable, and computer programming languages Unix and C. Keystone of his deft orchestration of many large-scale and diverse projects was encouraging interdisciplinary collaboration through the work environment and atmosphere.

Kelly believed in an "institute of creative technology," and this informed every decision from the architecture of the facilities to the type of people he hired and the location of their offices. The layout of the buildings at Bell Labs' headquarters, the Black Box designed by Eero Saarinen, allowed Kelly to introduce and enforce the policy of physical contact through forced proximity: a never-ending hallway with a literal *open-door* policy. Gertner describes the interdisciplinary interactions that occurred as "a physicist on his way to lunch…was like a magnet rolling past iron filings."

Another iteration of the Forced Encounters approach is Apple's under construction Infinite Loop headquarters: a Philip K. Dick-ian saucer-doughnut landed in the middle of Cupertino, California. The architecture is inspired by the infinite loop programming principle, which lists a sequence of instructions for the computer that continues endlessly. The floor plan of this ideologically eponymous building is similar to that employed by Bell Labs' vanishing point hallways, encouraging forced encounters between people.

Letter Together is an Indie typography workshop run by designers Jessica Hische and Erik Marinovich at their San Francisco studio, Title Case. During the events, designers draw letterforms, blending their own input within the intent of the whole: forced encounters on a personal scale. Hische invites twelve random participants across a one-day or two-day intensive period to collaborate on generating a refined, vectored alphabet by the end of the weekend. Dynamo doughnuts, coffee, fancy sandwiches, and beer appease new acquaintances and keep typographic spirits lively.

A second facet of Bell Labs' success was Kelly's orchestration of a diverse group of researchers in opposition to the prevailing trend of assembling homogeneous, *yessir* think-tankers. Groupthink, even under the IDEO form of Alex Osborn's Brainstorming, can eliminate aggressive outlier concepts by the peer pressure of critical mass, or worse, diminish personal responsibility within innovation. At Bell Labs, discussions were intentionally less top-down, favoring dialogue over consultation. A positive result of this interdisciplinary collaborative dynamic was smarter decisions supported by joint authorship.

A third factor was Mr. Kelly's belief in intellectual freedom as crucial to innovation, and he successfully managed to keep creativity separate from business, even if it meant spending lots of money on ideas that did not produce results for two years, if ever. By way of comparison, Facebook co-founder Mark Zuckerberg outlines one of Facebook's mottos as, "Move fast and break things." Bell Labs, on the other hand, followed a methodically collaborative code of deliberate open-endedness.

Taking cues from her former job at Google, Marissa Mayer began resuscitation of the flailing Yahoo! immediately after becoming Chief Executive in July 2012. Yahoo! had formerly been operating under the "Work From Home" 2.0 policy, but it determined isolation was not, in fact, working. Yahoo! struggled to maintain a relevant and consistent Internet presence. So in May of 2013, under Mayer's direction, Yahoo! switched things up and began renting four floors in the old *New York Times* building near Times Square. Now, 500 Yahoo! employees physically work amongst colleagues from Tumblr, in addition to independent tech firms 10gen and Citysearch.

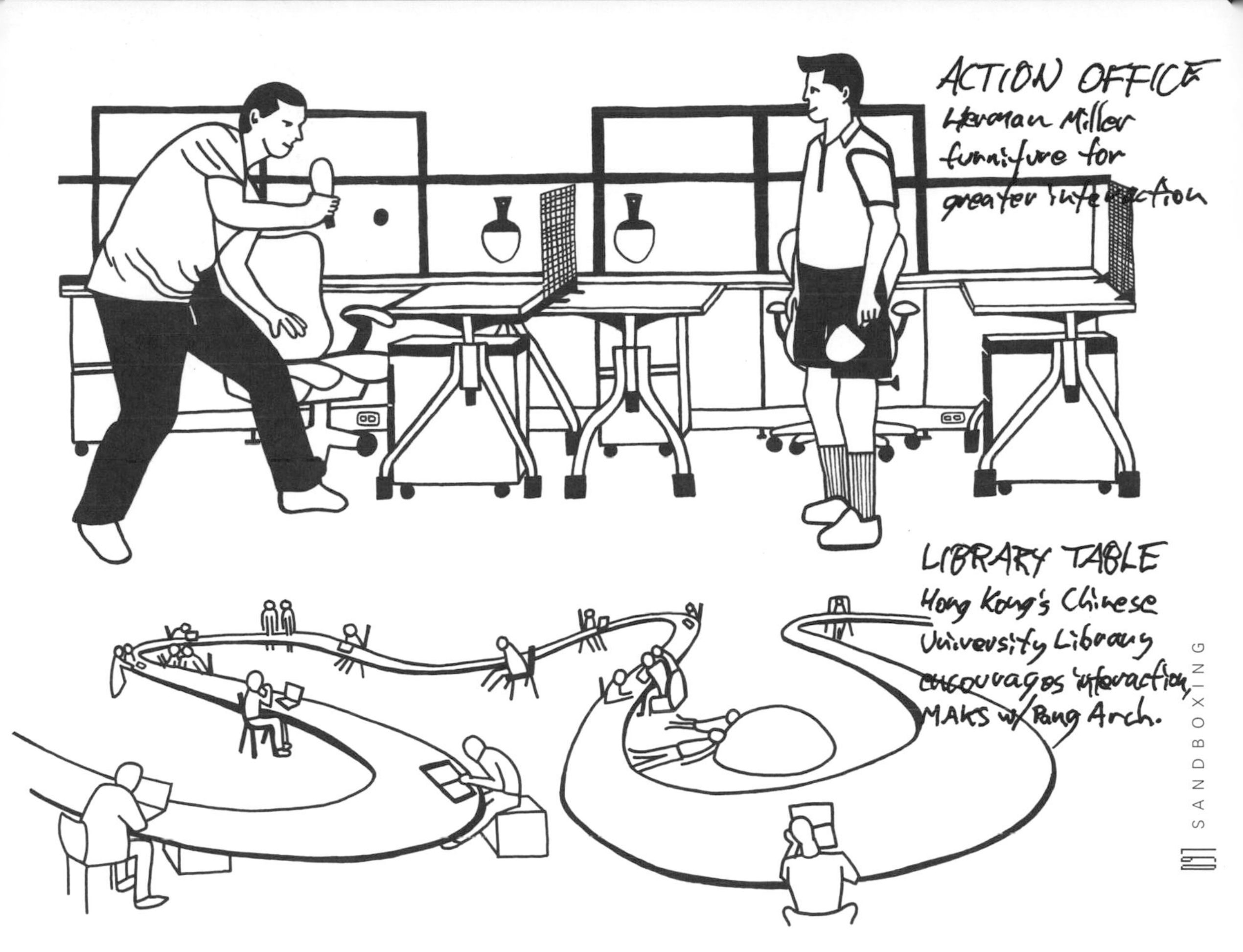

John Sullivan, a professor of management at San Francisco State University and the overseer of a human resources advisory firm, endorsed Mayer's push for collaboration as a means toward innovation. In *The New York Times* article, "Yahoo Orders Home Workers Back to the Office" by Claire Cain Miller and Catherine Rampell, Sullivan claims, "Studies show that people who work at home are significantly more productive but less innovative. If you want innovation, then you need interaction. If you want productivity, then you want people working from home." For graphic designers, production is important, but innovation is imperative.

One innovative environment example is Eyebeam, a smaller, non-profit, Indie warehouse that invites broad and diverse inventors, artists, designers, technologists, historians, writers, and anyone in between to conceive new media and advance culture through forced encounters, physical proximity, and open-ended production outcomes. Since opening in 1997, Eyebeam has conducted hundreds of fellowships, residencies, educational programs, workshops, marathons, performances, exhibitions, and lectures. Self-describing their mission, they say, "Eyebeam challenges convention, celebrates the hack, educates the next generation, encourages collaboration, freely offers its contributions to the community, and invites the public to share in a spirit of openness: open source, open content, and open distribution." For example, Data Visualization marathons hijack a couple hundred students nationally for a 24-hour collaboratively competitive workshop to best visualize contemporaneously significant data. For Eyebeam residents, innovation is directly tied to collaboration. Taken together, these examples indicate several helpful traits in constructing collaborative environments and atmospheres:

(1) Forced encounters between disciplines and practitioners

(2) Physical proximity of colleagues

(3) Open-ended production outcomes and deadlines

"It may be that we have become so feckless as a people that we no longer care how things do work, but only what kind of quick, easy outer impression they give." -Jane Jacobs

Building 20 + *Company*

The building: one concrete slab supported five wings of three floors constructed from a hodgepodge of large wooden posts, Masonite, gypsum wallboard, and tar paper as the cherry on top. Microwave radar was invented here. The building was the most inventive facility in America. Architects, shudder.

Building 20, the MIT eyesore in Cambridge, Massachusetts, was hastily constructed in 1943 as a makeshift Rad Lab, staging hundreds of scientists specifically solicited to develop a radar device that could identify distant German bombers during WWII. Given the harried circumstances at the time, people were considerably unconcerned about sporting a fashionable workspace. The result was a shoddy structure that, to everyone's surprise, stood fifty-five years longer than intended, and by the time of its demise, Building 20 had earned a reputation as a *magical incubator.* About a quarter of all the physicists in America had worked there at some point and nine of them won a Nobel Prize. How could a big ramshackle shed loom so large?

Mainly, because nobody cared about Building 20. It was cheap, expendable, under-designed, and temporary. The resident intellectuals felt at liberty to openly abuse the space and adapt it towards their own research needs.

Students and educators re-routed electrical wiring, suspended heavy machinery, and even installed anechoic chambers for testing sound quality. Jerrold Zacharias removed two floors in his lab to make room for a three-story atomic clock. The results were unpredictable, incessantly evolving. Building 20 provided the perfect collision of space and lawlessness to facilitate experimentation.

Intellectual diversity was the dominant feature. Physicists, linguists, model railroad club enthusiasts, hackers, sound engineers, electronics technicians, video game developers, anthropologists, philosophers, and the ROTC populated the grab bag. Urban theorist Jane Jacobs refers to this as *knowledge spillovers:* an exchange of ideas among individuals placed together. Jacobs says, "The proximity of firms from different industries affect how well knowledge travels among firms to facilitate innovation and growth." In other words, reducing the physical constraints of an environment staging an interdisciplinary group encourages cross-pollination of ideas, resulting in invention.

Amar Bose, incidentally, built the country's first hi-fi speakers while distracted by the Acoustics Lab down the hall. Morris Halle, in conjunction with Noam and Carol Chomsky, innovated syntax and phonology in two dingy offices. Even a homeless botanist squatting in a storeroom didn't feel out of place.

This type of interaction was partly spurred by the horizontal configuration of the space and a very poor labeling system. Rooms were not organized by subject or purpose, and chance encounters were normal. Unmovable equipment accumulated, and residents constantly adapted and repurposed space. This kind of unintentional remixing impacted by the environment is unlikely to occur in the hyper-specific designs, hierarchies, and organization of modern schools.

For contrast, Max Barry satirizes common corporate environments in his novel *Company*. Unlike the beautiful mosh-pit of intellects at Building 20, Barry's corporate building is overtly hierarchical. The floors of the high-rise office are ranked by importance, so employees can literally work their way up to the top. Level 1, which is the top-most floor, is for CEOs; Level 2: Senior Management; Level 3: Human Resources; all the way down to Level 19: The Call Center; and Level 20: The Lobby. The intentional result of the rules and division of labor is that none of the employees can explain what the Company *does* at all.

(4) Environmental liberties through spatial and structural malleability

(5) Intellectual diversity

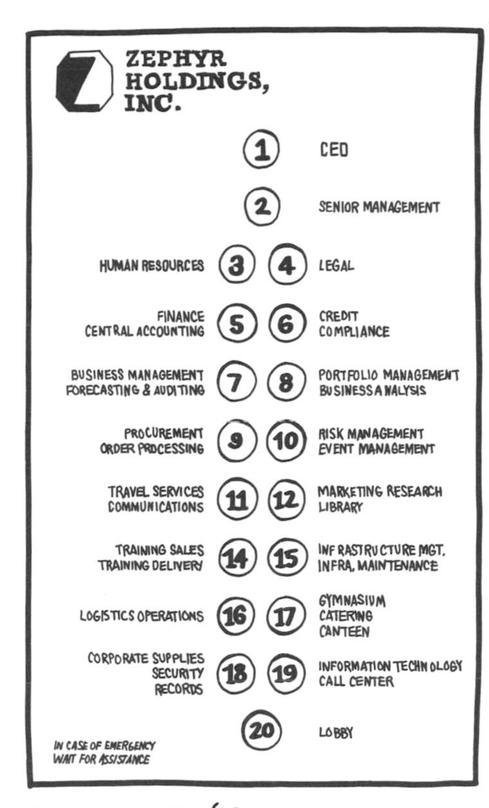

HIERARCHY
Max Barry's Company organized for less communication

TO PENNA. R.R. STATION

ADA LOUISE HUXTABLE
champion of humane public architecture

google circles

GOOGLE CIRCLES

GOOGLEPLEX

ROWBOAT ROOM
Googleplex collaborative creation space

Googleplex + Silver Factory + Appetite Engineers

What worked for Bell Labs and Building 20 also works for Danny and Sally in the Sandbox circa second grade. Collaboration functions best when the participants are free to try the unusual and the imaginative, playing with concepts to create a world. Grounding the narrative in the design community, compare the large-format brainstorming canonized at IDEO with the intrinsic differences of a small collaborative. Instead of outlier concepts getting swarmed under or held back altogether, all concepts are accepted outliers. Extreme concepts fit in, though they may be reigned in eventually; but it's hard to go the other way, taking the mundane and making it more interesting. Thus, the Play Element, not only in the early stages where the Sandbox is a blank canvas of latent exploration but continuing even after a direction is chosen, allows designers to freely evolve formal and conceptual schemes. The goals and confluence of circumstances with playing in a Sandbox are so utterly straightforward to kids: *obviously* the mountain needs to have a tunnel so that *Micro Machines* can roll through…

Googleplex has a big, red, multi-person (collaborative) slide. Google's Workplace as Playground concept provides spaces for interactions, planned and unplanned, as well as distractions for thinkspace. Hundreds of niche environments populate the 2,000,000-square foot complex. Everything from graffiti-ridden subway cars to Stanley Kubric space pods, row boats equipped with a comforter, and hammocks surrounded by palm trees create a collaborative atmosphere in which employees want to work and spend their free time on campus. Pixar employs a similar environment, including Razor scooters for transit, creating easy, fast, fun physical proximity in a large facility.

For a contrasting example, tin-foil wallpaper, silver paint, and silver balloons drifting along the ceiling were all it took to create a collaborative play atmosphere on a smaller scale and budget for New York's most famous Pop artist. Salvaged from the street, a red couch became a popular crash spot for guests. Billy Name "designed" The Silver Factory, Andy Warhol's studio and the hippest hangout spot in New York. Everyone from Mick Jagger and Salvador Dalí to Truman Capote and Allen Ginsberg congregated there. "Warhol Superstars" labored day and night preparing screens, running prints, photographing the assembly line, and producing films. The non-stop glam party—frenetic interdisciplinary interactions and collaborations, mixed with focused production—defined The Silver Era (1962–1984) as Warhol's most productive period.

Excuses and Distractions are also an integral part of design collaboration, and both The Factory and Googleplex provide fertile space for interaction and minds to idle through such prompts as game tables at bars, cappuccinos from Internet cafés, libraries, restaurants, and studio space. A relaxed, distracted state is less self-censoring, allowing for more extreme ideation, whether alone or in dialogue. Many designers entertain several collaborative projects simultaneously, allowing the works to incubate and provide mental relief from each other. This enables cross-pollination and alleviates Designer's Block. Research from the University of Chicago regarding ambient noise levels indicates that the hum of a coffee shop boosts creativity; there is a brainbalance sweet spot of noise prompts versus focused quiet in which the mind is able to wander without losing focus. Inspired by multiple examples of successful groups of designers and other Creatives launching from coffee shops, the data-fed site *Coffitivity* pipes in coffee shop noises through an Internet radio stream and promotes the science of distracted creativity.

THE FACTORY
Andy Warhol's center for productive 'interaction

THE SCIENCE OF SLEEP
Michel Gondry's blur of real/imagined

FREE WI-FI
Josh Ellingson does
Edward Hopper for Wired

OFFICE OF PAUL
SAHRE
design genius
has collaborative space

Somewhat similar to Warhol, graphic designer Martin Venezky has a factory, only it's more like a warehouse, and it contains lots of *kipple.* Coined by Philip K. Dick, *kipple* is the result of entropy on *stuff.* Venezky's studio, Appetite Engineers, is highly production-oriented, and much of Venezky's work relies on a quantity of available elements to physically play with, often physically building up designs through collage. By having a wealth of materials, influences, and gear on hand, his studio minions can engage in collaborative Sandboxing, where minds on break or at play have room to build on the fringes, often in direct response to environmental variables. Outside of the visual world, play is an important element of making. Writers' rooms within the TV and film industries are often known for using in-house banter and live conversation as production-oriented play, with the dialoguers building trust via laughter-ridden friendships.

(6) Play element
(7) Distraction or excuse element
(8) Accessibility of gear, tools, supplies

The Bauhaus + Cranbrook Academy of Art

Schools are a microcosm. As collaborative Sandboxes, many design programs seek to prioritize the accessibility of gear and tools too. Schools are in a unique position to provide the Freedom to Fail as a component of learning through risk-taking, exploring limits, and collaboration.

The Bauhaus was one of the first modern design schools, and it remains the graphic design industry's prototypical model for education. It existed in the time between the World Wars, from 1919 to 1933, and moved from Weimar to Dessau and finally ended up in a Nazi-ridden Berlin, where the school was forced to close. Founded by architect Walter Gropius, the Bauhaus was originally inspired by the Arts and Crafts movement in Belgium. The school resulted from a merger of the Grand Ducal School of Arts & Crafts and the Weimar Academy of Fine Art. As a result, the Bauhaus supported the idea of creating a *gestalt,* a total work of art. Arts and Craftsman William Morris (unsuccessfully) advocated this idea of combining art and life fifty years before the Bauhaus in an attempt to break arrogant barriers between fine artists and craftsmen and to bring high-quality design to all socioeconomic classes. The only hiccup was that his well-designed, everyday goods were too expensive for everyday life. Gropius stated that the goal of the Bauhaus was to

unify art, craft, and technology on an affordable budget, and that the students would be trained as applied fine artists working in the industry. The Bauhaus offered everything that a contemporary art school does: architecture, graphic design, photography, printmaking, industrial design, textile art, painting, pottery, sculpture, and theater. This interdisciplinary program, and the accessibility of the diverse practitioners, gear, and ideas, is at the heart of the Bauhaus model.

Packaged with interdisciplinary experimentation was the Freedom to Fail. Exploring relationships and trying a range of mediums carries a guarantee of some busted stuff, but it also leads to new inquiries.

There are three main reasons why the Bauhaus was a successful Sandbox. (1) The school had copious facilities fraught with professors actively conducting their own research amongst their students. (2) Barriers between disciplines were down-played. (3) An interdisciplinary curriculum + curious and enthusiastic wandering bodies = peer-to-peer education. The Bauhaus was Building 20 as an academic model. When German intellectuals fled to the States during WW2, Black Mountain College in Asheville, North Carolina inherited Bauhaus faculty, leading to scenes of Josef Albers on one side of the desk and Ray Johnson on the other.

Another oft-copied design school platform is Cranbrook Academy of Art in Bloomfield Hills, Michigan, a graduate art school influenced by interdisciplinary Bauhaus formats. Designers, architects, ceramicists, fabric designers, metalsmiths, painters, photographers, printmakers, and sculptors cross-pollinate on pretty much everything. At Cranbrook, students primarily evolve their work through crit as opposed to coursework—there are no *classes.* The informal structure also encourages open-ended experimentation, accepting that not everything will work. From 1971 through 1995, Katherine and Michael McCoy co-chaired the 2D and 3D MFA Design programs respectively, and revamped the Cranbrook agenda with more theory, experimentation, collaboration, and a *de*-emphasis on deadlines, papers, and finals. Faculty function as Artists in Residence, working alongside the students. It's true that the design work generated at Cranbrook tends to downplay the interests of the profession and align more with *design as fine art,* but this environment equips work that then defines the profession. Cranbrook has cemented its experimental status through contemporary rockstars, including Andrew Blauvelt, Ed Fella, Meredith Davis, and Martin Venezky.

A suite of gorgeous architecture creates a protected hive of making-based research set amongst a bombed-out Detroit. With ample studio space, beautiful grounds, a renowned museum, practicing faculty, and access to diverse mediums, gear, and perspectives, Cranbrook's inspiring setting encourages collaborative investigation and communication as both a means and ends.

(9) Safety zone and freedom to fail

(10) Inspiration and motivation

TABLE ASTERISK
raste on collaborative
type design

Calvin & Hobbes +
The Silmarillion

Calvin and Hobbes are relevant Sandbox bookends because imagination is Calvin's collaborative playground. Likewise, design is all about inventing worlds, with brand books being an overt example. Prior to writing *The Hobbit* and *The Lord of the Rings,* J. R. R. Tolkien started by developing a fully realized map of Middle Earth and even invented entire Elvish languages replete with dialects *(The Silmarillion).* When it came time to write his fantastical stories, the narrative felt real because the world *was* real. A similar logic applies to creating a collaborative work environment that enables the group to assume its own identity, resolved and separate from its parts.

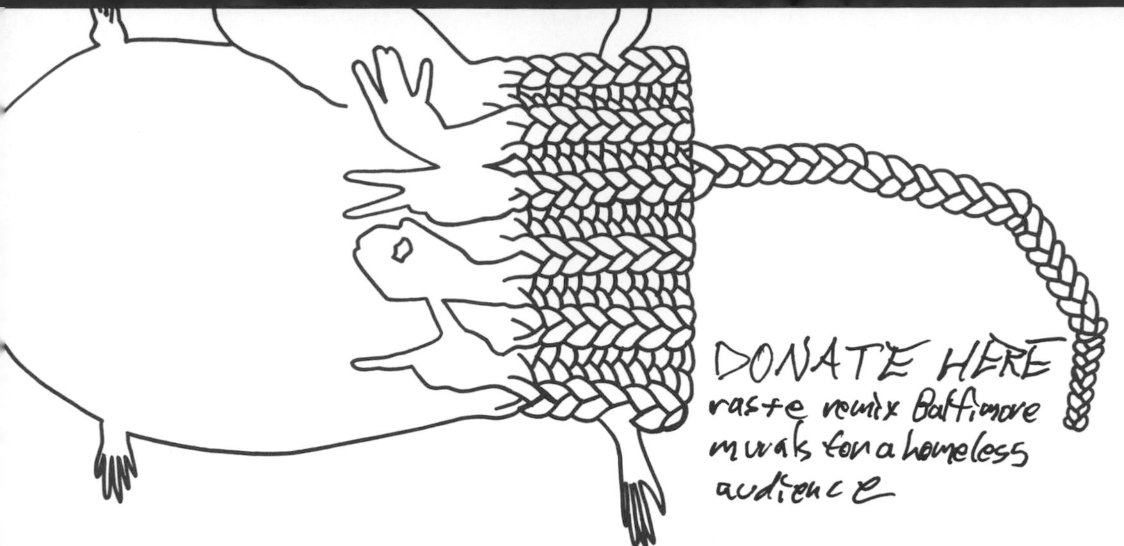

I hate it here—Yesterday, here in the middle of the City,
I saw a wolf turn into a Russian ex-gymnast and hand over a business card that read YOUR OWN PERSONAL TRANSHUMAN SECURITY WHORE! STERILIZED INNARDS! ACCEPTS ALL CREDIT CARDS to a large man who wore trained attack cancers on his face and possessed seventy-five indentured Komodo Dragons instead of legs.
And they had sex. Right in front of me. And six of the Komodo Dragons spat napalm on my new shoes.

Now listen. I'm told I'm a FAMOUS JOURNALIST these days. I'm told the five years I spent away from the City have vanished like the name of the guy you picked up last night, and that it's like I never left. (I was driven away, let me remind you, by things like Sickness, Hate and The Death of Truth.)

So why do I have to put up with this shabby crap on my front doorstep? Now my beautiful new apartment stinks of wet fur and burning dragon spit, and I think one of the cancers mated with the doormat. It keeps cursing at me in a thick Mexican accent. I may have to have it shot.

If you loved me, you'd all kill yourselves today.

—Spider Jerusalem

Transmetropolitan / Volume 2 / Warren Ellis

"People interest me, conversations don't."
-Hugh Laurie's character on House M.D.

Aggressive content presented aggressively pinned the poplove ambassador with a flood of inquiries....

Tagged as emerging artists, ras+e presented work during a gallery event with another local, but more established, Maker. The audience included a few fans, a few gallery execs from the region, and some vaguely curious folks drawn by the promise of free beer and hummus. Talks were short, with an emphasis on dialogue between the specifically paired/themed artists, the critics, and the community members. With an eye toward mentorship and interaction over "Look at what I made!" the street-designers avoided the standard Art Talk posturing, but the work was still the work.

The first presenter was known for spreading mural Love around the city. Whereas, ras+e showed a house-sized tech-for-eyes comic, which was a lot more dystopic than Love. The first artist's presentation featured repetitive imagery and sporadically superfluous detail, with long silences for the images to *speak for themselves*: "...and here's me and a stranger hugging at the grand opening of the Mulberry Street installation." In turn, the feedback the first presenter received was agreeable, uncritical, mushy, disengaged, and flat. The small group formed a numb posse at best, enablers at worst.

Presenter Two launched with background inspiration: a photograph of an Adult Video sign adjacent to a billboard reading, "Jesus is watching you." The aggression in the presentation along with the questioning posed through the lens of socially poignant work made the following Q&A engaging, informative, and investable for everyone. Some of this was design fundamentals: subverting expectations, interesting hooks, and curating language, all aimed toward an overt objective: meaningful dialogue.

For all the projects designers claim start conversations or promote awareness, the idea of designing language as a means to communicate effectively gets shrugged off as "content." But words matter, language matters, and effective communication is really just Design as Design. The core concept of image and text can be revised to include spoken language. Writing as process, at the level of collaboration, becomes conversation as process.

All the things that good design is and does can be found in engaged conversation.

Dialogue as Process: Design and Discourse

Ideally, language connects people. Language allows a culture to communicate complex ideas, exchange higher-level thinking, and express the subtleties of human emotion. It is also the most *exercised* aspect of any culture. Technically speaking, a language is comprised of a system of signs or symbols that have been ascribed to meanings, often arbitrarily. This enables the sharing of knowledge in the form of transactive memory, or collective consciousness. Thus, language has the inherent ability to appease autonomy, placate loneliness, and promote relationships, community, and social networks. Users give a language its value by agreeing on meanings, but it is its evolution and adaptations that make language powerfully active. Language is a web, highly specific in its parameters and elements, linking, holding, defining—the umbrella communication form. This is why the systems-thinking of precisely articulated *visual language* is an important design buzz term. Graphic designers are defined as visual and verbal communicators, and while this definition applies to all of the arts, no painters were overheard talking about *being communicators* during research for this book.

Designers regularly come up with smart slogans and rhetorical puns, obsess over their hierarchy and presentation, and even author essays and lectures, but seldom do we parse the everyday language used behind the scenes: how do designers communicate during critiques and project development, or charettes and conferences, or via email and messaging? Despite fetishizing formal process, the writing end of it can get overlooked. But all good writing is poetic, coming to the designer embedded with siphon-able rhythm and contrast. The fine arts have a highly formalized, long-standing tradition of language, criticism, and written theory with no equal parallel in design. Perhaps some of the design doctorate programs will help build this discourse, but for now a PhD almost ensures the writing and writer will be locked up in academia, separated from the Makers and making of the discipline. Designers tend to shy away from addressing our verbal and written process, including the methods in which we converse, critique, and communicate about work. Design discourse matters because the way designers talk affects the kinds of things designers are prone to think about and the level at which we're engaged to think about them.

I WISH FOR FUNNER CONVERSATIONS

SOCIALECT
social class dialect allows groups to bond via speech

I like your use of *line* in this piece.

Uh, thanks man.
I *value* you as a person.

Speaking hard truths, developing inventive and specific language to move collaborators through process and ideas, crafting "spark plug" moments through visual and verbal language, and exploring multiple approaches to the critique and conceptualizing stages are all important. Design school pushes for all these elements by emphasizing Process as Learning, and hiring faculty specifically to innovate and push students within Dialogue as Process.

Take _____ for instance. _____ says, "I like Sally's poster," and repeats this single idea over the course of 43 seconds, struggling with searchable synonyms for "like" (thanks, Facebook). Blushing like a roman candle, _____ concludes with, "You know what I mean." _____ is why critique faculty employ Exile as a teaching tool. While the class is thoroughly annoyed with _____, it's not entirely her fault. Sure, some intellectual curiosity about Tolkien might help; but her professor spec'd a mind-numbing design vocabulary book, her roommate watches too much of The View and listens too little to Eminem, her Twitter profile is HottiePants451, and her girlfriend uses emoticons "for clarity." The problem has less to do with a working knowledge of the words "symmetry" and "scale," than that a narrowing cultural language allows for narrow human expression. _____'s language sets her up for failure. She knows how to be "nice" and articulate "nice" because "nice" has the least amount of resistance and commitment. The umbrella-descriptors dilute communication so that _____ avoids scuffles; "nice" and "like" are the language of complacency, vagueness, and neutrality. Some _____s even promote themselves as Dumb, or a Blonde, or a Dude, or a Peon—a veiled, blanket OK to be boring or stupid. Instead of requiring a book of vague art terms, _____ needs *Slaughterhouse Five.* Oscar Wilde is an imperative before _____ gets to *Stop Stealing Sheep.* For _____ to meaningfully contribute to a conversation, a crit, writing a headline, or a process dialogue of any sort, _____ needs to understand and value nuanced language!!!!!

The ensuing breakdown of well-crafted language ties directly to design collaboration, placing ideas in cultural contexts. Dialogue is the most influential factor in smart interaction; designers can shapeshift, design, and control their vocabulary. Smarter language leads to smarter collaboration. William Isaacs, founder of the Dialogue Project at MIT, writes in *Dialogue: The Art of Thinking Together:* "[To] change the way we talk is to change the way we think. We can influence and regenerate the inner ecology of human beings by transforming the quality of our conversations." Designing effective collaboration requires designing effective dialogue, and the payoff is a well defined and highly specific means of production, as well as the ability to deliberately shift culture forward.

BROKEN BEE
neutering language

Unfortunately, economics and a declining valuation of design education and exploration are cutting into innovation, by knocking the legs out from under Process in favor of Speed and Money. Renowned interdisciplinary studio innovator Stefan Sagmeister noted as much in a *designboom* interview: "We have a group called 'Second Tuesday' and we meet every second month. There are about 15 people who run design firms. We always meet at someone's home or studio. That person has to organize dinner and a subject. Sometimes these subjects are quite practical such as finances. Lately the topics have been focused more towards administration and business rather than cultural aspects."

Dialogue as Process: Monotone Monolanguage

Society evolves language (sociolinguistics); conversely language directly impacts society (sociology of language). The cyclical result is that any cultural problems regarding language and design are exasperated. For collaborators and creators, context is everything.

ONLINE GALLERY
why studio courses will always be on campus

In America, semiotics have been increasingly hammered by modern mediums. The need for smart messaging rapidly declined in the '80s when a glut of cable TV meant sudden reams of unparse-able content. Form and concept development were downplayed, as flashing phosphor-dot fixes hypnotized society and proliferated communication, but of poorer quality. The situation got worse in the '90s with the rise in interactive media and the allure of perceived productivity and connectivity. It devolved further in the new millennium with the rise of cheap and accessible tech advertising in parallel to personal megaphonesocialsites, an "All-Speak" at the roller rink. Because Americans indiscriminately accept new technology, language and culture become piñata afterthoughts, valued in retrospect after they break. Artists, Makers, and Thinkers across a broad range of disciplines noted the results: Ted Kaczynski (theorist), Aldous Huxley (author), Neil Postman (culture critic), Corey Doctorow (tech writer), Kalle Lasn (activist), Brian Wood (comic writer), Pink Floyd (musician), and Nam June Paik (video artist). Oddly, designers, arguably the Makers most impacted by and responsible for the shifts, lack the same dedicated commenters of other disciplines. Postman labeled America the world's only *technopoly,* a critique

heyyyyy man
imho & fwiw ur
wrk sux

lol y?

Tweak Monotone

Monotone Control

Intensity:

Interesting Pitch — Flat Pitch

Reset

I don't like your tone

of its unquestioning absorption of technology-impacted culture, and Huxley feared that our love of uncritiqued tech would shift our society into irrelevance. Our gadget-lust, our data-worship, and our aimless digitized relationships are the American culture, mindlessly replacing real meaning, language, and interaction with a technological simulacrum. Guy Debord called it a *Society of the Spectacle:* "All that once was directly lived has become mere representation." Even Neo from *The Matrix* couldn't keep his Makers conversationally engaged.

Knowledge and conversation of any depth are ignored in favor of TED as education, priority given to the sleek and immediate bite-sized. We agreed on new meanings and crafted a sleek, quickfire language that flattened out dialogue, which led to a nuanced conveyance of human emotion and ideas. If "you only live once" was a hollow cliche, "#yolo" is Jello in a shallow babytalk pool. "Lol" has made lying conveniently automatic, while "hearts on sleeves" now feels more like the stuff of horror-schlock.

In short, language broke. Dialogue degenerated. An argument is that this is simply adaptation, that the shift is neutral. But the lack of specificity and humanity in design classroom discussion is stark and counterproductive.

dia•logue: (dia means through, logue means words or meaning), flow of meaning

con•verse: (con means turn, verser means together), to take turns speaking

dis•cussion: To shake apart, as in concussion or percussion

MAKE IT FASTER

brevity	abv
bad	sux
great	gr8
photo	pix
high contrast	hic
download	suck dwn
search	google
dear	heyyyyy
impatience	u thre?
!	!!!!!!!
tone	:) :/ :(

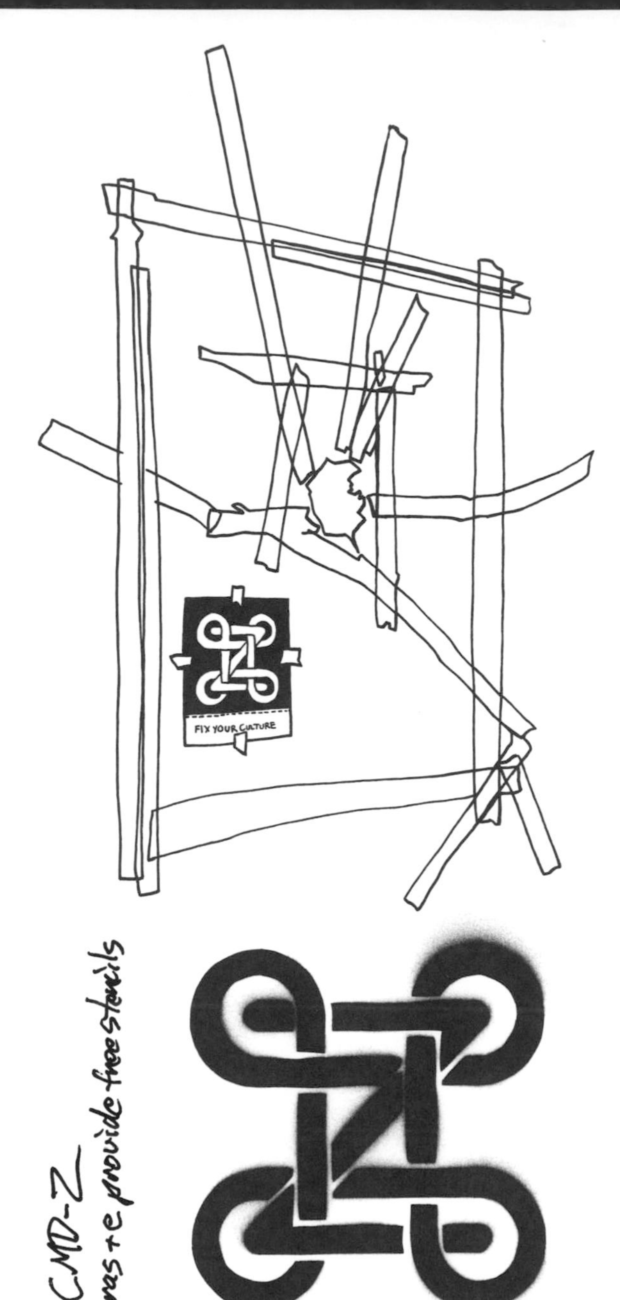

"And it spreads. Rupert Murdoch beams his shit into Asia, English children are taught that Z is pronounced Zee by goddamn Barney, and all of a sudden, world cultures become the Monoculture, the same conversation, the same clothes, the same show. All tuned to Channel Zero."—Introduction by Warren Ellis from Channel Zero by Brian Wood

Many elements created the problem. Social media networks contributed to a pandemic of cleaned up, politically correct mono-language as companies took ownership of user data and personal communications were crafted via checkboxes. In the '90s, hoards of people contributed to sensationalist social media driven by quick-views, first via television then Internet, creating the viral mass hysteria of Americans' Culture War on language. Political correctness was introduced to our lexicon in the '50s, dulling language, but the rise of prosumer tech, accessibility of personal commentary, and instant publication of thought created a new atmosphere of non-speak. There are two major, relevant results: everyone saw precisely how rough unfiltered internal language could be as masks were peeled back by the meanness of blunt language stripped for speed, murdering nuance and friends. Whether conditioned by fear wrought by a sensationalist mainstream media, or because the uncontrolled flood of new, poorly constructed content and language absorbed room for thought, the Politically Corrected became thought/language police, while their polar opposite Trolls reduced all discourse to yelling. The "intellectually disabled" America still celebrates "Winter Holiday" with a monstrously garish tree in Rockefeller Center. Political correctness makes language timid and unspecific, the opposite of increasing diversity. The veneer of additional syllables doesn't fool anyone. Auto-corrected language sounds as bad as auto-tuned singing, and anyone with something meaningful to say, or sing, zippers shut in fear. Our flattened culture no longer wants to speak, a problem for anyone interested in collaboration or making: nobody wants to produce neutral, toothless work.

At the same time, like taking speed to offset the drowsiness of alcohol, digital social networks have encouraged a blanket informality. Humor + language reduces to the briefest, select from the options, memes. Learning abilities atrophy to 140-character bursts or 18-minute lectures, while lengthier investigation of thought sends the brain, twitching, into stasis. Texting abbreviations have infiltrated everything from letters, emails, and even everyday speech. Retweets have replaced journalism. Emoticons have become so accepted that professionals are using them in the workplace. Judith Newman's on-point *New York Times* article, "If You're Happy and You Know It, Must I Know, Too?" details how the crutch of universal

tribute to Paul Rand
"Writing is a matter of
having ideas."

PARALANGUAGE
nonlexical component
of communication

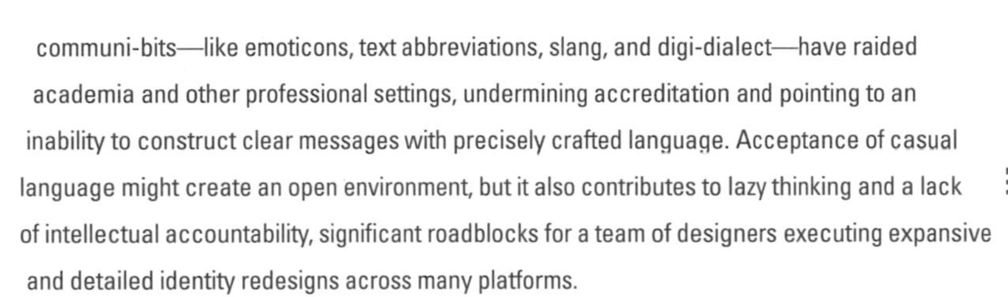

communi-bits—like emoticons, text abbreviations, slang, and digi-dialect—have raided academia and other professional settings, undermining accreditation and pointing to an inability to construct clear messages with precisely crafted language. Acceptance of casual language might create an open environment, but it also contributes to lazy thinking and a lack of intellectual accountability, significant roadblocks for a team of designers executing expansive and detailed identity redesigns across many platforms.

LET'S TALK BUSINESS!!!!
contemporary forms of communication have altered what is considered professional conversation

In an interview with Miggs Burroughs for *Miggs B. on TV,* Paul Rand remarked, "Writing is a matter of having ideas. Most people write well, but there's no substance to what they write, so it's pretty boring stuff, even if the adjectives and nouns and pronouns fit. But that's not the point. The point is that the ideas have to fit. And you know, writing has something to do with music. It has to sound right. And we think in words. We think, 'Today I'm going out, tomorrow I'm not going out, and now I'm gonna sleep and next week I'm gonna do this.' And if you think thoughts that make sense, you can write them down. And you can be a good writer." Expanding this, quality thinking produces good design, good communication, and meaningful collaboration.

The flattening of language is a global flattening. Comics legend Warren Ellis writes in a '90s rant in the Introduction for Brian Wood's *Channel Zero:* "And it spreads. Rupert Murdoch beams his shit into Asia, English children are taught that Z is pronounced Zee by goddamn Barney, and all of a sudden, world cultures become the Monoculture, the same conversation, the same clothes, the same show. All tuned to Channel Zero."

In both *pragmatics* and *semiotics,* the new "American Language" is the biggest influence on our societal, and global, inability to think and speak critically. This severely undermines the desire to collaborate and the process of doing so.

"The most important work in the new economy is creating conversations."
-Alan Webber, Editor of *Fast Company*

Dialogue as Process: Thinking Together

Fruitful collaborative conversation seems esoteric, but it's an acquirable skill—designers must be spry—and it demands practice. Dialogue is a shared experience that expands and combines individual memories, allowing original ideas and new perspectives to build from a blended space. Dialogue is group-sketching. It demands recognition of your own voice and process, while acknowledging how it works within a group. An individual may fill different rolls, or their vision may have a different impact, from group to group. Dialogue is experimentation, testing reactions out loud; it's live, immediate, and if the ball drops, the game is over. Just like design is completed by the viewer or user, Collaborative Designers need to account for intended and unintended responses to their verbal actions, and they must learn how to close the gaps between misperception and personal motive.

In William Isaacs' *Dialogue: And the Art of Thinking Together,* he notes four elements of dialogue: listening, respect, suspension (taking turns), and voice. When all of these are controlled, the chances for productive conversation increase. Exercising individual cognition obviously improves collaborative conversation. Reading extensively, having an informed and holistic background across the arts, staying abreast of contemporary culture and events, and general self-awareness are requisites of any graphic designer. In addition, there are some social tenets that designers should consider for improving their conversational game. Our cues come from social psychologists, linguists, anthropologists, and businessmen, but also from authors, comedians, political leaders, directors, musicians, performance artists, UX/UI designers, and philosophers: all Thinkers and Makers contribute to constructing idea-worlds.

Organizing Genius authors Warren Bennis and Patricia Ward Biederman build a case for the creation of internal language as an indicator of successful creative group work. Inventing a world reinforces and defines the group members and their work but, more importantly, it allows Creatives to precisely define objectives, while encouraging further creative thought not beholden to strictures outside of the world.

language

by George Carlin

post traumatic stress disorder = shell shock
sneakers = running shoes
information = directory assistance
the dump = the landfill
motels = motor lodges
used cars = previously owned vehicles
riot = civil disorder
strike = job action
jungle = rain forest
glasses = prescription eyewear
garage = parking structure
drug addiction = substance abuse
prostitute = sex worker
theater = performing arts center
wife beating = domestic violence
constipation = occasional irregularity
broke = negative cash flow
kill = neutralize
cripples = passengers in need of special assistance

How to Design Your Dialogue (DYD):

1. Free Speech

+ Do not censor or inhibit your ideas or reactions.
+ Disregard expected social mannerisms.
+ Do not be ashamed.
+ Do not confuse people for their ideas.
+ Lose track of time.

These elements center around a core collaborative design concept iterated throughout the CO LAB ideology: the objective is a synthesis of the inventors who transcend themselves as parts; addition—A+B≠C—is not enough; rather, synthesis demands a result that is entirely new beyond its components. A unique entity—A+B=Curious George—is unexpectedly created from a collision transcending an assembly line's distribution of labor approach. A + B + C + D + E + F + G + H + I + J + K + L = Magazine + N + O + P + Q + R + Spread. The Free Speech concepts demand a blend of uncritical thought mixed with critical dialogue to spur open-ended group-process evolution. Old school Brainstorming is an ineffective generation principle because The Mass inevitably, though unintentionally, pushes outlying—highly inventive—ideas to the margins, either because tacit consensus forms around a nucleus of ideas as discussion runs on a thread or contributors self-consciously avoid speaking about oddities.

All this amiability must give ground to human emotion, interaction, prolonged play, and idea-faucets left wastefully open for prolonged periods. No ideas should be attacked or ignored; competition in a collaborative should always be from a desire to explore deeper, and contributors should not defend beyond a general Socratic sense. Encouraging an internal world reinforces "We're all in this together." Acknowledging "Us Against the World" allows designers to feel like anything can be tried because there's nothing to lose. All craziness will be considered for potential; then, atmosphere and interactions are designed to provoke and encourage.

These interactions necessitate engagement. Designers are sometimes too willing to emotionally detach from the content, process, or product, draining energy and inertia from Deep Process. The reality of open-ended exploration is that it may require an open-ended time frame. Reality often means deadlines, and the design process thrives on speed, but young ideas sometimes need time to grow up: design dialogue to nurture all

rough concepts without forcing early maturation and while watching out for the bully of easily coalesced critical mass.

Example: *"Life is more than money / Time was never money / Time was never cash"*— Switchfoot in "Gone"

Example: "People in each section of the ballroom tended to stare at the nearest voice-box, instead of watching the distant figure of whoever was actually talking far up front, on the podium. This 1935 style of speaker placement totally depersonalized the room. There was something ominous and authoritarian about it. Whoever set up that sound system was probably some kind of Sheriff's auxiliary technician on leave from a drive-in theater in Muskogee, Oklahoma, where the management couldn't afford individual car speakers and relied on ten huge horns, mounted on telephone poles in the parking area. ...Their sound system looked like something Ulysses S. Grant might have triggered up to address his troops during the Siege of Vicksburg."—*Fear and Loathing in Las Vegas: A Savage Journey to the Heart of the American Dream* by Hunter S. Thompson

SPEECH PATTERN

2. Substance: No Bullshit

+ Eliminate jargon that alienates outsiders.
+ Eliminate academic speak that obfuscates meaning.
+ Push past expected, scripted, cookie-cutter responses.
+ Push past euphemistic ("soft") language.
+ Push past politically correct language.

Example: "Am I a genius? I don't think so. Not yet anyway. As Burt would put it, mocking the euphemisms of educational jargon, I'm exceptional—a democratic term used to avoid the damning labels of gifted and deprived (which used to mean bright and retarded) and as soon as exceptional begins to mean anything to anyone they'll change it. The idea seems to be: use an expression only as long as it doesn't mean anything to anybody. Exceptional refers to both ends of the spectrum, so all my life I've been exceptional."—*Flowers for Algernon* by Daniel Keyes

Example: "You with your big words, and your, your small, difficult words."—Peter Griffin, *Family Guy*

Education has not had a great recent track record with bullshit-less dialogue: consumerist reasons at the

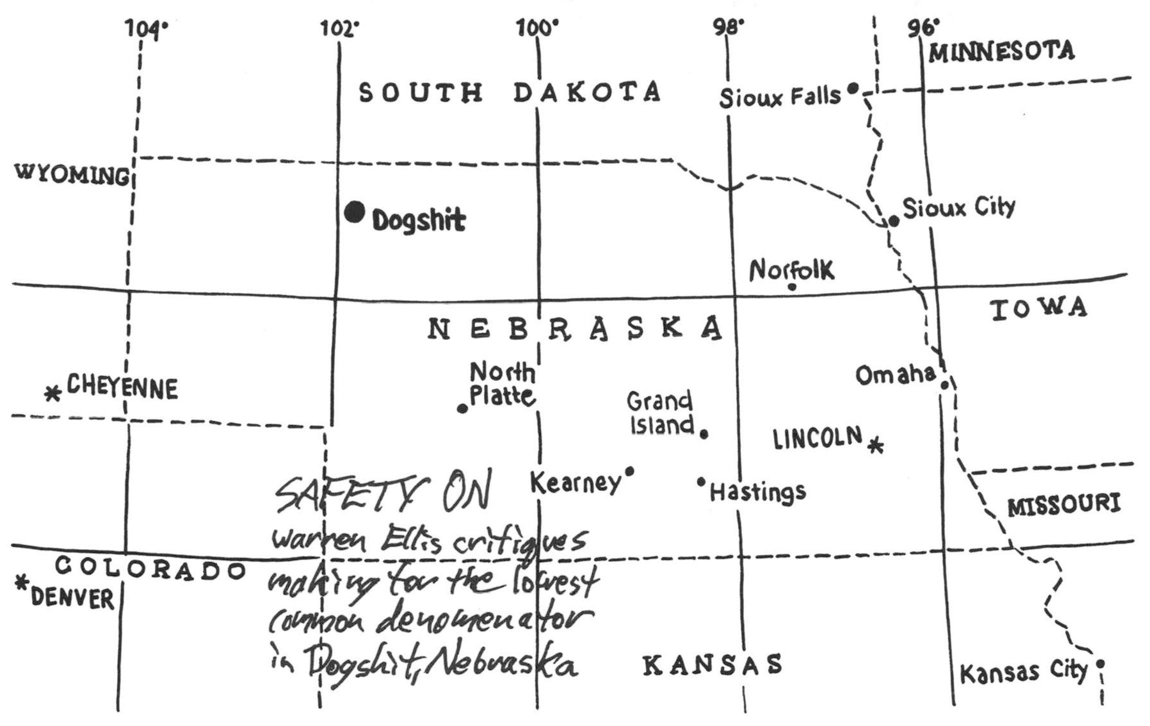

college level and pro-testing/anti-teacher climate in grade schools. In either environment, dialogue suffers because language suffers because content suffers. Colleges have seen a spate of Spring 2014 graduation speakers—former UC Berkeley boss Robert Birgeneau, Condoleeza Rice, IMF head Christine Lagarde—forced out by student protesters who found their contrarian views inappropriate for a celebration of academic progress.

In his article "The Wilds of Education," *New York Times* columnist Frank Bruni pointedly asks, "When it comes to bullying, to sexual assault, to gun violence, we want and need our schools to be as safe as possible. But when it comes to learning, shouldn't they be dangerous?" Several Koch-backed takeovers of school boards banning books indicates that the best way to learn about essential life experiences is by not discussing them, or at least not discussing them critically.

Designers chafe when clients shy away from aggressive and impactful work; eliminating these traits in education ensures the lack of such work in the future professional landscape, while preventing young practitioners from developing self-confidence in self-aware processes.

Good design needs imps. This means education needs precise vocabulary, rampant curiosity, and dangerous dialogue in a safely collaborative environment.

Design faces the distinct challenge of creating punchy and meaningful work, as corporate BLAH can drive invention toward the safest common denominator. Comics writer Warren Ellis prizes the punch in his foreword to Brian Wood's *Channel Zero:* "[Comics] don't have huge corporations trembling at our every movement, because we make no money compared to the other visual narrative media. That vast commercial pressure isn't brought to bear on comics. Which means, often, that we can say what we want without rich men's scissors attacking our work until it's safe for little Tommy in Dogshit, Nebraska. I hate little Tommy in Dogshit, Nebraska. I want to kill little Tommy in Dogshit, Nebraska." The Koch brothers exactly do "leverage commercial pressure" to attack language and literature and discussion in schools because they tremble at young collaboratives—and schoolchildren—discussing accurate and interesting means of representing their empire.

VERBOSITY
my ears hurt

3. Salience: No Fragmentation

+ Be direct and accessible.
+ Be accurate and precise.
+ Be concise, not verbose.
+ Do not oversimplify.
+ Use engaging vocabulary and references.

These all sound more like writing tips than design tips, but dialogue and design and all other forms of communication follow similar principles. Our cultural critics—like Stephen Fry, Jon Stewart, Stephen Colbert, Cory Doctorow, Neil Postman, Patton Oswalt—are essentially writers. Their status in comedy, acting, activism, and comics all reduce down to their interest, training, and capabilities in wordcraft. Translation: for designers to truly impact culture positively and collaborate effectively, they need to understand, care for, and master language in all its forms.

Examples: "The Angry Mob" by Kaiser Chiefs vs. "The Metropolis and Mental Life" by Georg Simmel

"We are the angry mob
We read the papers every day
We like who we like
We hate who we hate
But we're also easily swayed"

vs.

"The deepest problems of modern life derive from the claim of the individual to preserve the autonomy and individuality of his existence in the face of overwhelming social forces, of historical heritage, of external culture, and of the technique of life. The fight with nature which primitive man has to wage for his *bodily* existence attains in this modern form its latest transformation. The eighteenth century called upon man to free himself of all the historical bonds in the state and in religion, in morals and in economics. Man's nature, originally good and common to all, should develop unhampered. In addition to more liberty, the nineteenth century demanded the functional specialization of man and his work; this specialization makes one individual incomparable to another, and each of them indispensable to the highest possible extent. However, this specialization makes each man the more directly dependent upon the supplementary activities of all others. Nietzsche sees the full development of the individual conditioned by the most ruthless struggle of individuals; socialism believes in the suppression of all competition for the same reason. Be that as it may, in all these positions the same basic motive is at work: the person resists being leveled down and worn out by a social-technological mechanism. An inquiry into the inner meaning of specifically modern life and its products, into the soul of the cultural body, so to speak, must seek to solve the equation which structures like the metropolis set up between the individual and the super-individual contents of life. Such an inquiry must answer the question of how the personality accommodates itself in the adjustments to external forces."

ACTIVE VOICE
versus wussy voice

4. Active Voice

+ Stagnant and stale language leads to stereotypical design.
+ Craft sentences deliberately.
+ Do not be boring.
("All God does is watch us and kill us when we get boring. We must never, ever be boring."—*Invisible Monsters* by Chuck Palahniuk)
+ Use words that have energy.
+ Use active (not passive) verbs: DO vs. tell.
+ Phonetic language (speech sounds), rhythm, and pacing matter.

Example: "Good evening, London. Allow me first to apologise for this interruption. I do, like many of you, appreciate the comforts of everyday routine—the security of the familiar, the tranquility of repetition. I enjoy them as much as any bloke. But in the spirit of commemoration, thereby those important events of the past usually associated with someone's death or the end of some awful bloody struggle are celebrated with a nice holiday, I thought we could mark this November the 5th, a day that is sadly no longer remembered, by taking some time out of our daily lives to sit down and have a little chat. There are, of course, those who do not want us to speak. I suspect even now, orders are being shouted into telephones, and men with guns will soon be on their way. Why? Because while the truncheon may be used in lieu of conversation, words will always retain their power. Words offer the means to meaning, and for those who will listen, the enunciation of truth. And the truth is, there is something terribly wrong with this country. ..."—*V for Vendetta,* directed by James McTeigue, based on the book by Alan Moore

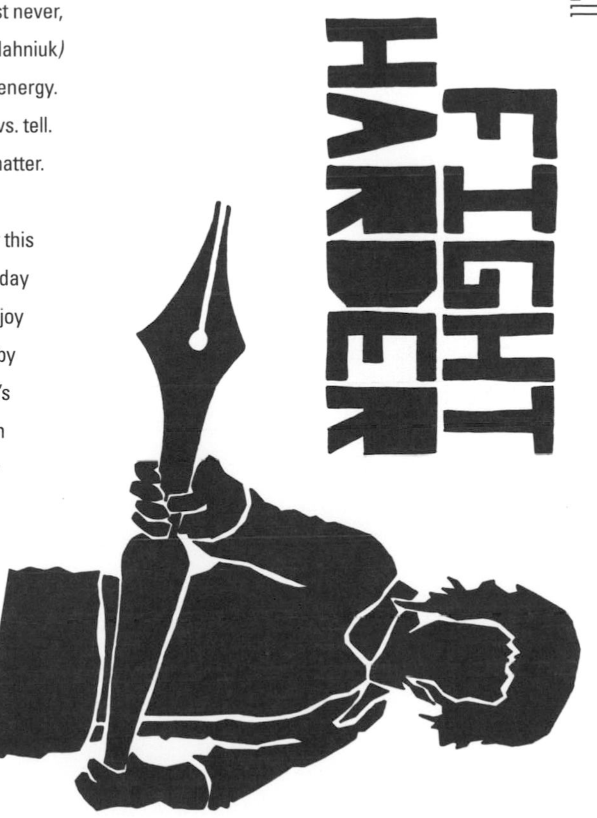

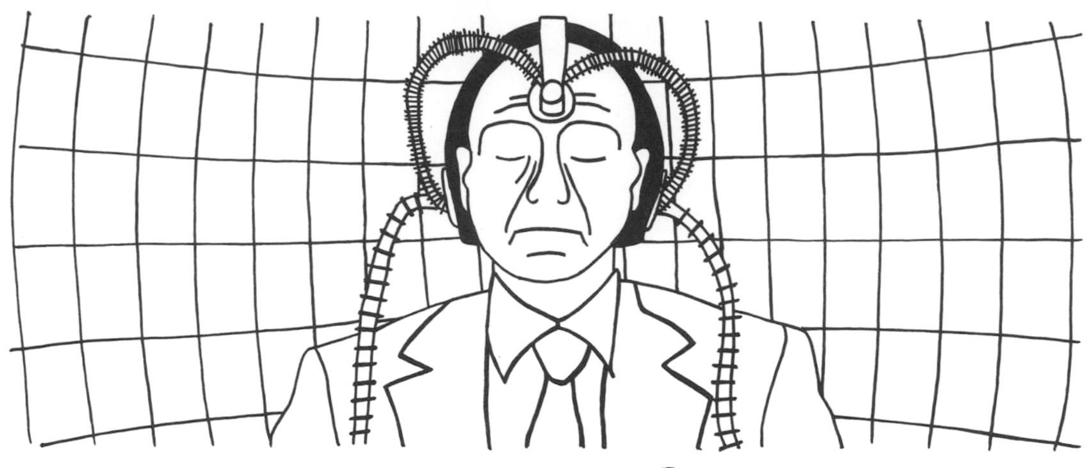

CEREBRO
X-MEN knew how to
connect with people

5. Socialect (Social Class Dialects)

Restricted code is a speech pattern that facilitates strong bonds, solidarity, and pride amongst lower-class group members. Conversely, *elaborated code* is used by middle and upper classes to gain access to higher education and professional advancement. Bonding is not as strong and speech tends to be fake and unemotional. More emphasis is placed on "I" than "we."

Use regional speech patterns (dialect, accent, slang, acronyms) to your advantage. That is why verbalizing text messaging abbreviations in oral conversation sounds idiotic, because the context has shifted. However, irony culture enables out-of-context speech, as it does everything else.

marco/POLO
DISTINCTIVE
LANGUAGE
people can be "read" and
identified by their speech

Example: *Jules:* Fuck, nigga, what the fuck did you do to his towel?
Vincent: I was dryin' my hands.
Jules: You're supposed to *wash* 'em first!
Vincent: You *watched* me wash 'em.
Jules: I *watched* you get 'em wet.
Vincent: I was washing 'em. But this shit's hard to get off. Maybe if I had Lava or something, I coulda done a better job.
Jules: I used the *same* fuckin' soap you did and when I got finished, the towel didn't look like no goddamn *Maxi-Pad!*
—*Pulp Fiction* by Quentin Tarantino

Example: "I mean, I am a writer, I deal in words. No, there is no word that should stay in word jail, every word is completely free. There is no word that is worse than another word. It's all language, it's all communication. And if I was doing what you're saying, I'd be lying. I'd be throwing in a word to get an effect. And well, you do that all the time, you throw in a word to get a laugh, and you throw in this word to get an effect too, that happens, but it's all organic. It's never a situation where that's not what they would say, but I'm going to have them say it because it's gonna be shocking. You used the example of "nigger." In *Pulp Fiction*, nigger is said a bunch of different times by a bunch of different people and it's meant differently each time. It's all about the context in which it's used. George Carlin does a whole routine about that, you know. When Richard Pryor and Eddie Murphy do their stand-up acts, and say nigger, you're never offended because they're niggers. You know what they're fucking talking about. You know the context in which it's coming from. The way Samuel Jackson says nigger in *Pulp Fiction* is not the way Eric Stoltz says it, is not the way Ving Rhames says it. They're all coming from different places. That word means something different depending on who's saying it."—Quentin Tarantino in an interview with Erik Bauer for *Creative Screenwriting Magazine*

6. Speech Stripped of Status

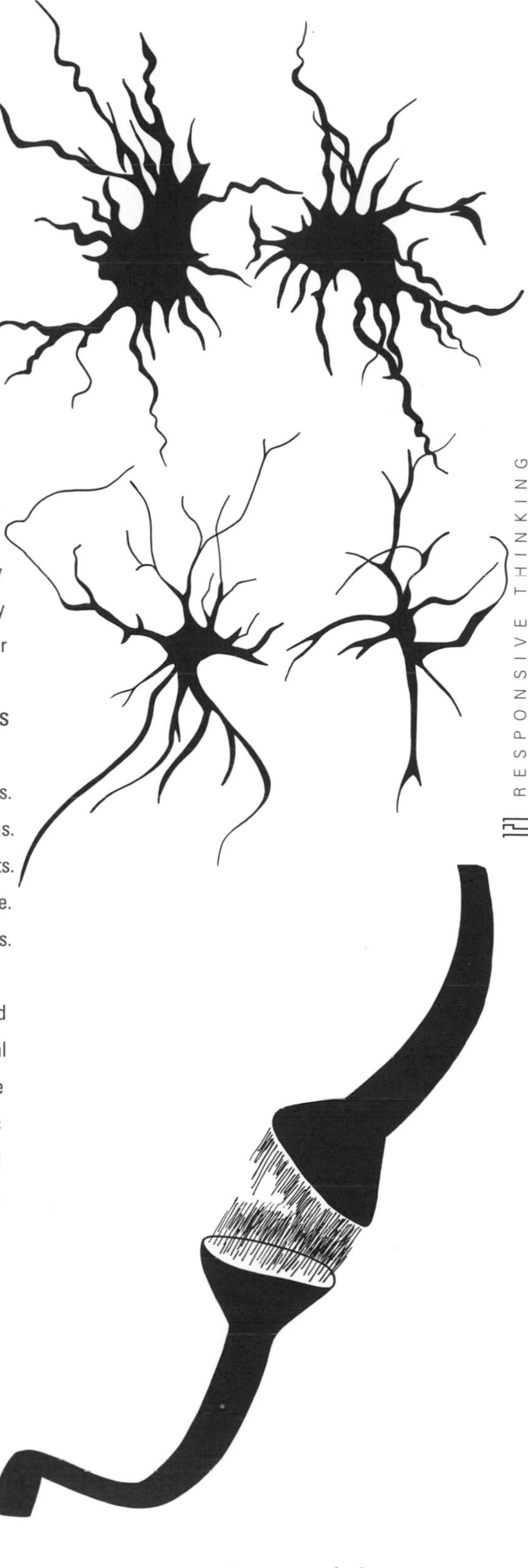

+ Remove language that inflects professional qualifications/insecurities.
+ Balance formal and informal, professional and friendly tendencies.
+ Learn delegation without becoming a despot.
+ Respect one another on equal terms.

There is a balance that all good conversationalists strike between distinguishing themselves without pushing others to feeling like outsiders. Personalities never displace others. They are instantly relatable but never chameleons, who are almost always outsiders. Targeting language toward an audience without giving up personal traits is a skill. J. D. Salinger's Holden Caulfield commented on when that skill becomes phony.

Example: "I am always saying 'Glad to've met you' to somebody I'm not at all glad I met. If you want to stay alive, you have to say that stuff, though."—*Catcher in the Rye* by J. D. Salinger

7. Forced Connections

+ Connect and jump between disparate ideas.
+ Anything and everything can be used as a reference for ideas.
+ Allow logic trains to develop by pursuing tangents.
+ Allow the unexpected to shock and surprise.
+ Avoid dialogue/logic loops.

Example: "He got on stage in front of hundreds of fellow economists and sang an a cappella version of what he and his friends call 'The Social Capitalist Theme Song.' Without warning, he starts to sing: 'The more we get together, the happier we'll be. Because your friends are my friends and my friends are your friends. The more...' He stops mid-sentence and focuses on me from behind his metal-rimmed glasses. 'If the audience just sits there like you are, grinning, then I stop. And I say, 'You don't get it.' The whole point is not about me singing to you. It's not about being amused. It's not about being entertained. It's about us singing the song together. It's doing things together that makes us happy.' Back on the #4 Powell, I realize that talking to the gangster knitter didn't just make me happier, it probably made her happier as well."—*Meme Wars: The Creative Destruction of Neoclassical Economics* by Kalle Lasn

Example: "Write the book the way it should be written, then give it to somebody to put in the commas and shit. I can't allow what we learned in English composition to disrupt the sound and rhythm of the narrative. If it sounds like writing, I rewrite it."—Elmore Leonard

NEURONS
collaborate at Godspeed

8. Suspend vs. Defend

+ Be inquisitive and skeptical. Blind faith is lazy.
+ Debate is healthy. Defensiveness is annoying.
+ Advocacy is a positive alternative to defensiveness.
+ Speak *with* as opposed to *at.*
+ Be open-minded and flexible to change.
+ Be goal-oriented.

Those last two seem like a paradox: keeping a goal in focus provides motivation, prevents burnout, and can motivate investment in the crazy tangents that lead to immaculate solutions. Because if it was easy and obvious then nobody would need designers. Chuck Klosterman is a good interviewer because he comes with write-able concepts in mind that he allows to bounce off his subjects without getting in their way.

Example: "To get a more specific example, I ask him [Kilmer] about the 'toll' that he felt while making the 1993 *Western Tombstone.* He begins telling me about things that tangibly happened to Doc. Holliday. I say, 'No, no, you must have misunderstood me—I want to know about the toll it took on you.' He says 'I know, I'm talking about those feelings.' And this is the conversation as follows:

CK: You mean you think you literally had the same experience as Doc Holliday?

Kilmer: Oh, sure. It's not like I believed that I actually shot somebody, but I absolutely know what it feels like to pull the trigger and take someone's life.

CK: So you're saying you understand how it feels to shoot someone as much as a person who has actually committed a murder?

Kilmer: I understand it more. It's an actor's job. ..."

—"Crazy Things Seem Normal, Normal Things Seem Crazy," *IV* by Chuck Klosterman

Example: "Psychologically and socially, Great Groups are very different from mundane ones. Intrinsically motivated, for the most part, the people in them are buoyed by the joy of problem solving. Focused on a fascinating project, they are oblivious to the nettles of working together in ordinary circumstances."

—"The End of the Great Man," *Organizing Genius* by Warren Bennis and Patricia Ward Biederman

9. Gauge Temperature and Emotional Control

+ Learn to empathize and forgive.
+ Keep personal investment in check.
+ Anger and frustration are detrimental.
+ Passion is more effective than credentials or talent.
+ Invest in other people.
+ Keep group morale high with periodic encouragement.

Example: "I don't want to be at the mercy of my emotions. I want to use them, to enjoy them, and to dominate them." —*The Picture of Dorian Gray* by Oscar Wilde

Example: "...anybody's life is valid, you know. But to really get to know people and discover humanity, which is what I truly think writers and actors do, you've got to be interested in other human beings, you have to be interested in humanity in general, and you have to do some discovering of humanity and different people. In real life there are no bad guys. Everybody just has their own perspective. I do have sympathy for the devil. To keep pursuing that you need to break out of your environment, whether that is Hollywood or you're a novelist living in Rhode Island. You gotta go have a conversation with and get to know somebody that makes $10,000 a year. You know, they have a different fucking perspective. So that's the only danger, you've gotta work at it, you gotta work at going out and keeping your hand into other people's lives and not just your own."—Quentin Tarantino in an interview with Erik Bauer for *Screenwriting Magazine*

10. Paralanguage

+ Hearing tone, inflection, and emphasis clarifies meaning.
+ Grammatical gymnastics make conversations more lively.
+ Adopt personas that facilitate dialogue.

Donelson Forsyth describes *paralanguage* in *Our Social World* as the auditory equivalent of body language. Watching someone's eyes as a lie detector test has equivalents in vocal twitches. More subtly, we can hear the things people don't say, the adjustments of breath, stammers and mumbles, and strengthening projection. These cues have no dictionary meaning, but we intuitively read into what we hear.

Example: "A philosophical question: from which angle to start looking at life, god, ideas, or anything else. Everything we look at is false. I don't think the relative result is any more important than the choice of patisserie or cherries for dessert. The way people have of looking hurriedly at things from the opposite point of view, so as to impose their opinions indirectly, is called dialectic, in other words, heads I win and tails you lose, dressed up to look scholarly. If I shout:

Ideal, Ideal, Ideal

Knowledge, Knowledge, Knowledge

Boomboom, Boomboom, Boomboom

I have recorded fairly accurately Progress, Law, Morals, and all the other magnificent qualities that various very intelligent people have discussed in so many books in order, finally, to say that even so everyone has danced according to his own personal boomboom, and that he's right about his boomboom: the satisfaction of unhealthy curiosity; private bell-ringing for inexplicable needs; bath; pecuniary difficulties; a stomach with repercussions on to life; the authority of the mystical baton formulated as the grand finale of a phantom orchestra with mute bows, lubricated by philtres with a basis of animal ammonia."— *Dada Manifesto 1918* by Tristian Tzara

11. Descriptive Language

+ Deploy words as if it were any other artistic medium.
+ Adopt a vibrant vocabulary informed by aggressive and diverse reading habits.
+ Ensure that adjectives and adverbs are not boring or obtuse.
+ Use examples, metaphors, and anecdotes.
+ Avoid details that do not give the piece something important.

Example: "The beet is the most intense of vegetables. The radish, admittedly, is more feverish, but the fire of the radish is a cold fire, the fire of discontent not of passion. Tomatoes are lusty enough, yet there runs through tomatoes an undercurrent of frivolity. Beets are deadly serious. Slavic peoples get their physical characteristics from potatoes, their smoldering inquietude from radishes, their seriousness from beets. The beet is the melancholy vegetable, the one most willing to suffer. You can't squeeze blood out of a turnip... The beet is the murderer returned to the scene of the crime. The beet is what happens when the cherry finishes with the carrot. The beet is the ancient ancestor of the autumn moon, bearded, buried, all but fossilized; the dark green sails of the grounded moon-boat stitched with veins of primordial plasma; the kite string that once connected the moon to the Earth now a muddy whisker drilling desperately for rubies. The beet was Rasputin's favorite vegetable. You could see it in his eyes."—*Jitterbug Perfume* by Tom Robbins

12. Graphic Design Vocabulary

The official ras+e list of words to kill:

simple
nice
clean
basically
minimal
like (why?)
cool, sick (so, not "good?")
good
dunno
Carson (as a basis for every comparison)
pretty (unless it's an insult)
style
human-centered (because, no shit)
subjective
optimization (wait, we want to see the tangents)
stuff
c'mon, y'know
whatevs (should probably be after "dunno")
auto-
troll (nix assholes, then we'll move it up to the next rung)
snapshot, pic, pix (anything that encourages people to treat photography irreverently)
for s(h)ure, fo sho, and totally (inflection included)
well...

YOU CAN'T SQUEEZE BLOOD FROM A TURNIP
'intro for Tom Robbins' Jitterbug Perfume

language is designed

Example: "A man with a scant vocabulary will almost certainly be a weak thinker. The richer and more copious one's vocabulary and the greater one's awareness of fine distinctions and subtle nuances of meaning, the more fertile and precise is likely to be one's thinking. Knowledge of things and knowledge of the words for them grow together. If you do not know the words, you can hardly know the thing."—*Thinking as a Science* by Henry Hazlitt

janky

simple

UBER!

EPIC

you know

you know

u know

basically

style

nice

sick

NEW ART JARGON
it's worsening

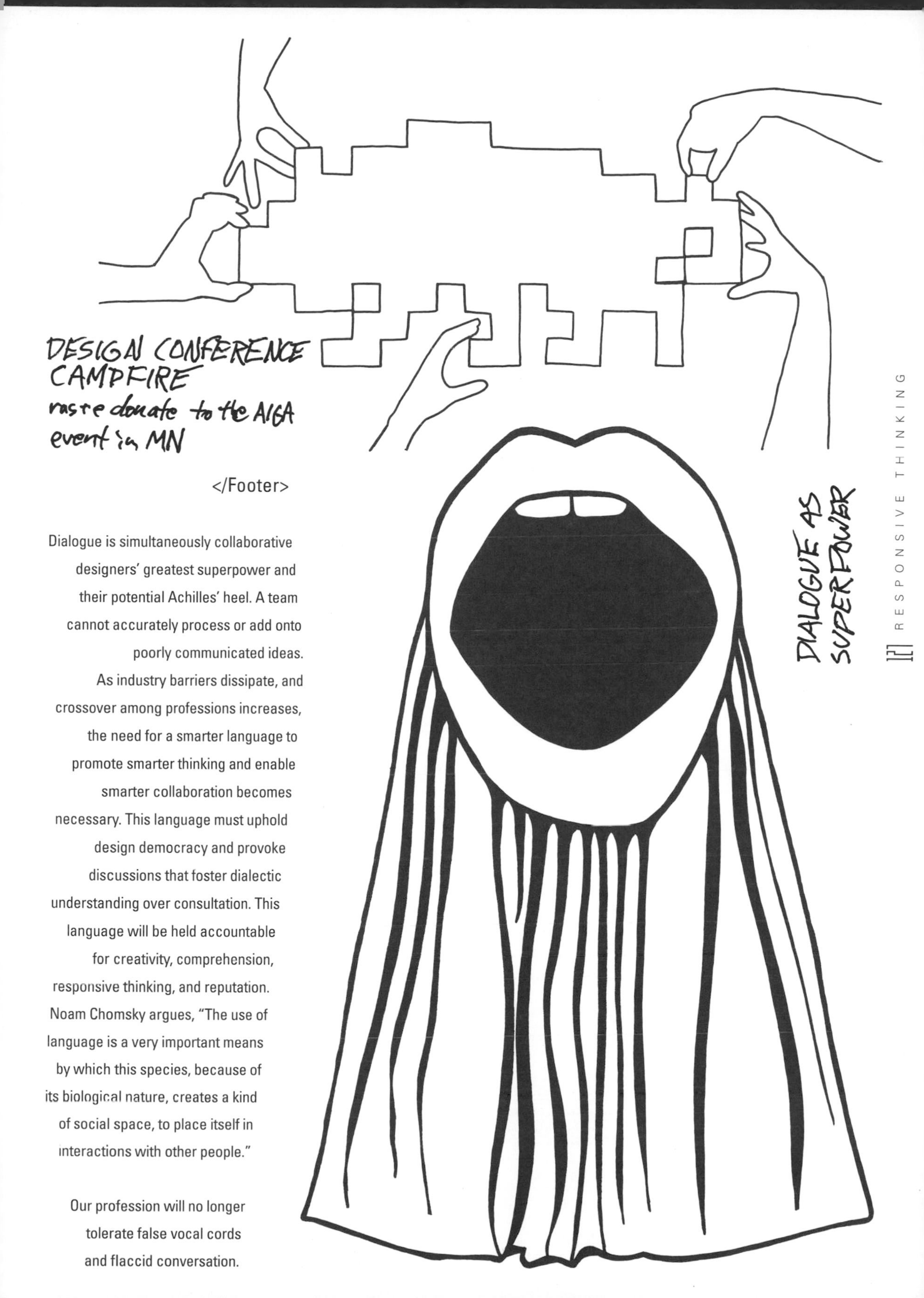

</Footer>

Dialogue is simultaneously collaborative designers' greatest superpower and their potential Achilles' heel. A team cannot accurately process or add onto poorly communicated ideas. As industry barriers dissipate, and crossover among professions increases, the need for a smarter language to promote smarter thinking and enable smarter collaboration becomes necessary. This language must uphold design democracy and provoke discussions that foster dialectic understanding over consultation. This language will be held accountable for creativity, comprehension, responsive thinking, and reputation. Noam Chomsky argues, "The use of language is a very important means by which this species, because of its biological nature, creates a kind of social space, to place itself in interactions with other people."

Our profession will no longer tolerate false vocal cords and flaccid conversation.

SIGHTSEEING

I don't get it

BUMP

Jane Kennedy

573-9421

KENNEDY@GMAIL

It looks better from here #siteseeing

"I can't believe NASA is getting rid of SpaceBook. We were on the verge of learning if social networking sites could support intelligent life!"-Stephen Colbert

After three hours of open studio time, my senior thesis students slowly trickled out of the classroom. I had spoken to a handful of them, reviewing research and project endeavors. The majority of the students were happy to have the time to work without interruption. Earbuds went in, and a fence of black-rimmed eyeglasses reflected a neutral backlit glare. Some students scattered into the lounge to work more comfortably with laptops in laps.

"Can I email you about my project later?" She asks this on her way out.

"Well, can we just talk now? What's up?" Said me.

She explained that she wanted to share some things she had been working on in class and when I asked her why it would be better to have this discussion via email rather than in person, she said that she didn't know, but that it was just easier for her to email it to me. I told her that she could, but if that was the case, then she probably should have been taking online courses instead, since physical edification wasn't helpful to her. She then told me that online classes were a joke and left, waving happily with errant flip-top mittens, and resumed a half-complete text message on her iPhone. Another student yelled after her, "Text me laters!"

A few factors could be at play here: (1) The act of writing an email helps this student practice composing her thoughts with an audience in mind, (2) She actually doesn't have anything new to share and is buying time, (3) Emailing is a more comfortable form of communication that removes the sense of immediate responsibility and the time constraints of physical interaction, (4) She prefers her online identity to her physical one, (5) Discussions via email are more real or normal, and I'm being abnormal by talking to her directly, or (6) She's lazy and just wants to leave now.

I would like to believe she is not lazy and likes writing, but based on her in-class proclivities and vocabulary, she simply prefers interacting remotely and sees in-person class and communication as inconveniences. This scenario, where people sign off with plans to continue a conversation online, regardless of actual follow-through, has become our culture's version of *goodbye*. But it's not *goodbye*. It's *I'm away* but "you're all still gonna be there when I need you, right?" A continuous conversation is known as a *perma-sation* and we're immune to noting it anymore. The idea of *co-presence* applies to physical separation but with omnipresent connection via technology. As a new generation of designers grows up with constant connectivity, and heavily relies on technology to mediate their interactions, what impact does this have on how humans relate to each other overall? Furthermore, how can we make technology better facilitate design collaboration? The truth is, our tools connect and our tools isolate. For designers crafting messages, and reliant on technology to do so, as well as aiming to collaborate with each other, parsing these mediums for the line of diminishing returns is an ethical and qualitative essential.

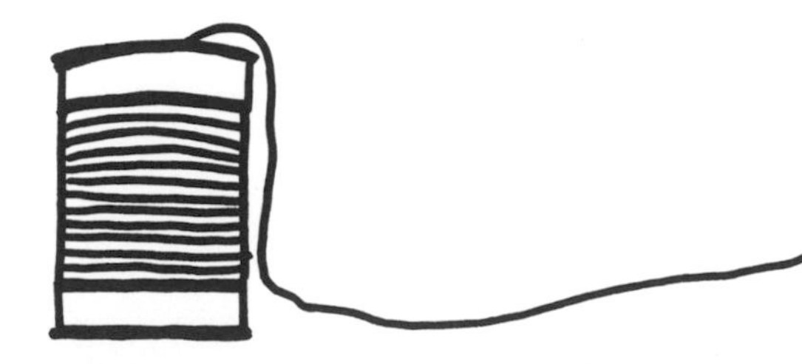

Technological Socialism

In a single day, we talk, text, chat, post, link, review, comment, email, fav, play, jam, upload, download, bid, crowdfund, listen, watch, edit, compete, hijack, remix, send, and apply. We respond to an onslaught of bombardments that tell us to HEY! PARTICIPATE! DONATE! CHOOSE! VOTE! SHARE! JOIN!!! BUY!!!! LIKE!!!!! Anymore, the majority of our socializing is mediated through contemporary communications mediums.

Interactive apparatuses are two-way communications devices that enable human-to-human (vs. human-to-computer) correspondence. Despite the friendliest of interfacial PHP, a machine can only humor humanity—judging, processing, and following an action script based upon input with third-party insensitivity. Interactive apparatuses can be anything from what we might expect, like cell phones and the Internet, but also include board games, sex toys, calling cards, laser pointers, dining and conference room tables, intercoms, walkie talkies, beepers, emergency help stations, the postal service, and mail. *Connectivity* refers to the proverbial wires that moderate our interactions with one another. More than ever, we have devices and systems (and excuses) that help us interact with each other in every imaginable relationship and scenario.

Part of the reason for this increase in connectivity has resulted from *prosumer* technology—the cross of *pro*fessional and con*sumer* markets resulting in affordable, semiprofessional gear. More people have more access to more quality equipment. Digital cameras, laptops, cloud storage, hosting and domain services, printers, projectors, and smart phones are affordable for the masses. As a result, the diverse boom in publishing options creates a range of rockstar *authorship* avenues. Lo-fi, democratic design equipment, in addition to a DIY perspective and open-source tools, has unfettered the field of graphic design. An elite toolkit and training defined graphic designers in the past; now we are sharing ideas and promoting work within a much larger and more diverse population. The design world has rapidly shifted from hierarchial, closed, compartmentalized states and individual endeavors to an open and global community.

Invitation + Access + Acceptance —> Participation

STAND OUT ONLINE. IT STARTS WITH THE RIGHT NAME.

99¢*.COM

PROSUMERISM

LSD APP

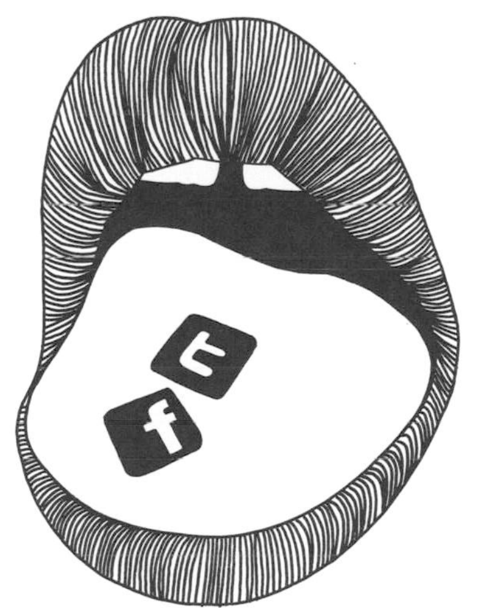

"The economic system founded on isolation is a circular production of isolation. The technology is based on isolation and the technical process isolates in turn. From the automobile to television, all the goods selected by the spectacular system are also weapons for constant reinforcement of the conditions of isolation of 'lonely crowds.'"
-*Society of the Spectacle* by Guy Debord

I HAVE A SIGNAL

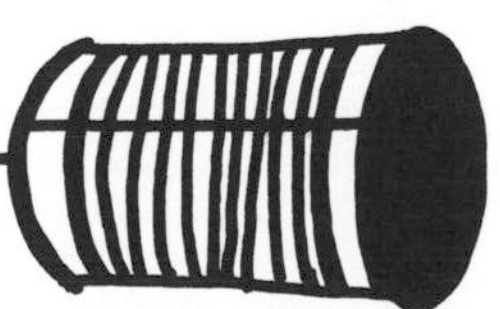

Furthermore, technology has enabled us to come together more quickly, easily, and specifically, based on special topics or interests instead of settling on who is available at the local bar. Collaborative membership is no longer tethered to money or geography. Technology has given us the opportunity to choose what we do and do not want to be a part of. Anymore, we are not predestined, captive users. We are voluntary, selective, vocal participants.

Technological socialism is a theory of social organization based on all of these components—the horizontal structures offered by technology. The Internet, case in point, is a decentralized network that lacks sovereign leadership. It promises equal accessibility and dissemination of information, and participants are peers with responsibility for decision-making, production, distribution, and exchange. Despite regular attempts by national governments and large corporations to exert control, the Internet is largely owned and regulated by the global community as a whole; voices, tools, and groups arise in response to opportunity or threat with immediacy. Anyone with an online presence is a citizen of this technological socialist society.

Historically, geographically based socialist organizations exploited and monetized the labor of collectivism. However, *digital socialism* can be thought of as "a cultural OS that elevates both the individual and the group at once," as defined by Kevin Kelly in the *Wired* article "The New Socialism: Global Collectivist Society Is Coming Online." "In the late '90s, activist, provocateur, and aging hippy John Barlow began calling this drift, somewhat tongue in cheek, 'dot-communism.' He defined it as a 'workforce composed entirely of free agents,' a decentralized gift or barter economy where there is no property and where technological architecture defines the political space." Kelly elaborates on the concept, arguing that these technologies amalgamate as a form of socialism despite lacking a formal document stating such: "When masses of people who own the means of production work toward a common goal and share their products in common, when they contribute labor without wages and enjoy the fruits free of charge, it's not unreasonable to call that socialism."

A similar perspective is echoed by Clay Shirky, a Professor of New Media at New York University studying the topology of social networks and how Internet technologies shape our culture and vice-versa. In *Here Comes Everybody,* Shirky maps the progression of participant involvement within groups. Putting the argument in the context of collaborative design culture: origins of technological socialism grow slowly as Makers learn and adapt.

1. Sharing. There has been an incredible willingness of people eager to present their lives for others, to the point where Web 2.0 sites (and beyond) are utilized primarily as diaries. The underlying phenomenon is not limited to these technologies—also functioning in physical manifestations of interactive design thinking—such as Candy Chang's *Before I Die* or Ji Lee's *The Bubble Project.* Such real-world sharing seems to circumvent concerns finally gaining traction in the U.S. regarding certain forms of unfiltered online diaries, namely that there is little payoff for all the self-opening.

2. Cooperation. Individuals come together to achieve a larger goal, such as Kickstarter, Wikipedia, or The Million Dollar Homepage. Sharing and promoting group knowledge, and creating a public library, has little of the diary's drawbacks—a point vociferously decried by corporate interests in cases like Napster, or anything involving Disney.

3. Collaboration. A step beyond cooperation, technological collaboration involves interaction, work, and groups based upon shared goals. There is usually no monetary reward. Says Shirky, "[T]he work-reward ratio is so out of kilter from a free-market perspective, workers do immense amounts of high-market value work without getting paid that it often makes no sense within capitalism. ...Of course there's nothing particularly socialistic about collaboration per se. But the tools of online collaboration support a communal style of production that shuns capitalistic investors and keeps ownership in the hands of the workers, and to some extent those of the consuming masses." Collaboration happens when the ends—when the group identity—become more important vehicles for change than the individual perspectives or contributions.

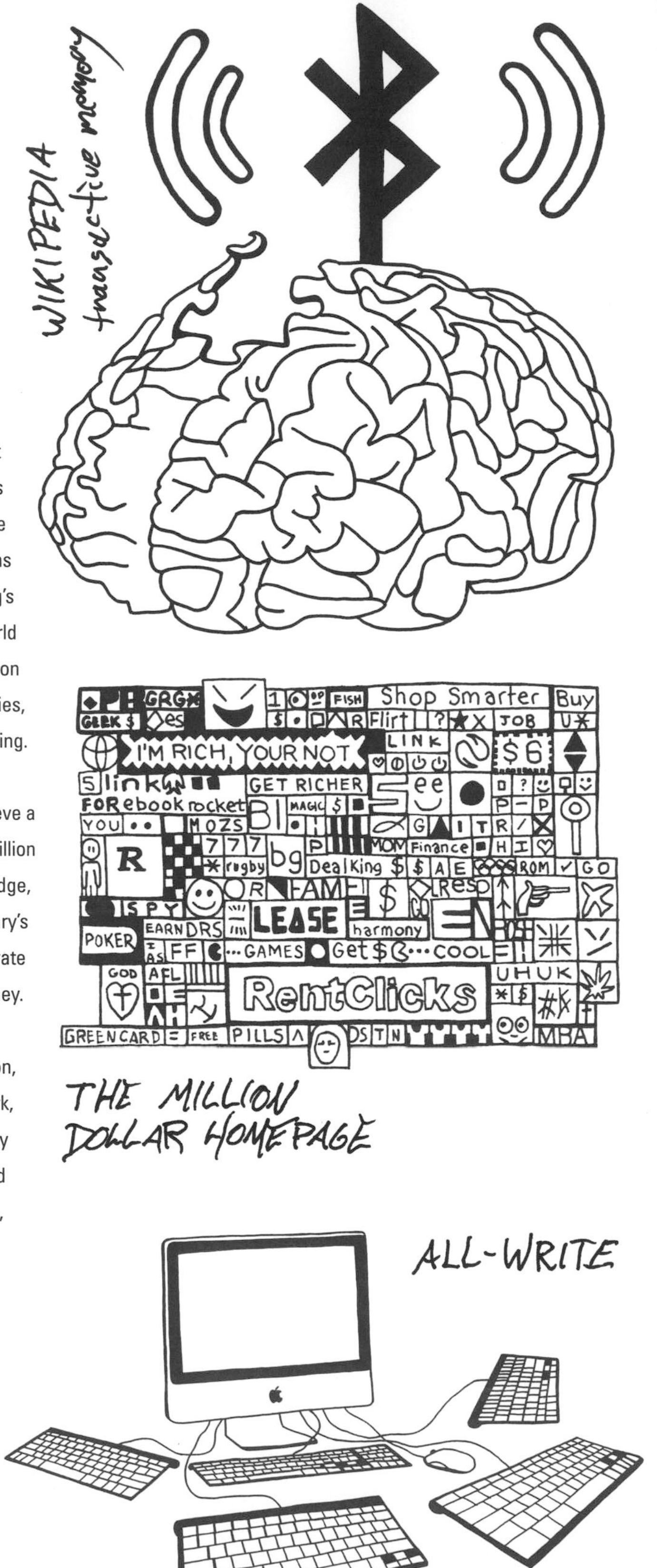

4. Collectivism. As group coordination, activity, and passion increases, collaborations can become cult-like. Shirky argues that this online collectivism functions as "stateless socialism," putting an emphasis on developing and defining a group beyond temporarily coming together for a specific project. Parsing what constitutes a collective vs. a collaborative sometimes reduces to investigating semantics, but at least in terms of historical examples of art collectives/collaboratives, the real difference comes between a group of individuals pursuing individual work under an agreed message, vs. a *whole* with a point of view specific to the group that extends beyond the contributions of individuals. In a collective, the parts maintain self-identity and purpose, making the results more of a loose assemblage of multiple threads, much like Shirky's political metaphor.

Graphic designers have always been on the forefront of the latest technology to both share and publish projects. We participate in the same, grand technologically socialist society as everyone, though we are historically keen on profiting from our participation in a way other participants are not. For many of us, social-hipness and tech savvy pay the bills and sustain roofs. On the spectrum of involvement, this places designers more cooperative than collaborative, or even collective.

Our ongoing argument: small design studios in a diverse range of physical locations are more searchable, socially spry, and able to compete with large ad agencies thanks to accessible, democratic communications technologies and an arsenal of linked-in, interdisciplinary designers. The resulting organization has less to do with pure efficiency—money—than designing an environment angled toward beliefs in collaboration, its process and specific results. Pure efficiency does not necessitate starvation; smaller studios are readily available to large businesses, and many projects come from local clients with a community-oriented mindset that values artistic responsibility and group interplay, all of which enables voice-heavy, decentralized studios. The combined result is a design democracy in need of concept-driven, idea-first authorship to rise above the competitive corporate masses of designers with singular interests or skills. These studios, with their shifted emphasis, encourage

CO CO CO CONNECTED

CALL ON ME

"When people live in an environment of such circuitry and feedback, carrying much greater quantities of information than any previous social scene, they develop something akin to what medical men call 'referred pain.' The impulse to get 'turned on' is a simple Pavlovian reflex felt by human beings in an environment of electric information. Such an environment is itself a phenomenon of self-amputation. Every new technological innovation is a literal amputation of ourselves in order that it may be amplified and manipulated for social power and action."-*War and Peace in the Global Village* by Marshall McLuhan

MAP OF COLLABORATION
Oliver H. Beauchesne at Science-Metrix (2005-2009)

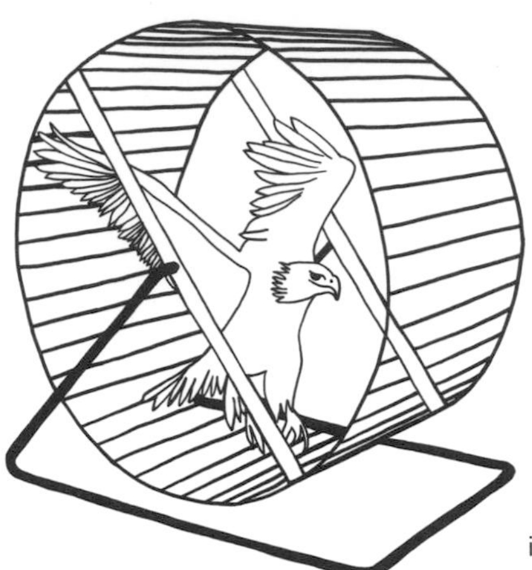

graphic designers to willingly explore technology in altruistic, collaborative ways. Technological socialism shrinks the footprint of specialized savants and eminent design sovereignty. Discussions tend to be less top-down and foster dialectic understanding over consultation. This socially humble, anti-profiteering agenda manifests in Linus Torvalds, founder of Linux; Peter Buchanan-Smith, founder of Best Made; and Tim Hoover with Ryans Smoker and Martin, founders of the Infantree—all of whom launch projects from a collaborative perspective built outside the demands of ruthless efficiency.

HOWARD ZINN chronicled the underclass of America on constant repeat

These efficiency demands are well documented in every industry but design, where innovation is as often about beauty, and creating a space where beauty can occur, rather than about quantity. Howard Zinn's seminal *A People's History of the United States* details with numbing repetition how Winners have always made their fortune and clout off the backs of the Losers. This includes hijacking indigenous people groups for local labor, abusing children as industrial cogs, and importing Chinese workers to form rail beds. Much of China's recent industry combines all these variables as rural children move into specialized zones for factory labor supporting American and European corporations. From that vantage, replacing even more human jobs with robots makes sense, though we seem to be doing so without considering how to support a populace whose labor is no longer needed. "Work" and "Jobs" will need to be rethought. Technological socialism will take on increased relevance in creating an all-winners structure. Culture critics and artists have investigated the possibilities, including author Corey Doctorow's hypothetical adhocracy based on bio-tech and social popularity scores in *Down and Out in the Magic Kingdom.*

Successful technologically social collaboratives, those in the here and now and those proposed, have a few key identifying traits:

1. **Active Participation.** A merger of invitation, access, and acceptance enables participation, but *active* involvement requires that collaboratives keep up and move forward with the initiative.

2. **Investment.** As a tech collaboration builds speed, members become addicted to their endeavor. This outputs as passion, pride, and red-eyed dedication. Invested participants are constantly connected and interacting.

3. **Decentralized.** Technological socialism is a horizontal hierarchy, with little or no governmental authority, and it is not tied to any geographic nationality. If there are minor leaders, they have little practical power and function more as symbols. This also means individual compensation and fame are not primary goals.

4. **Informed.** Self-education opportunities abound, marked by *Choose Your Own Adventure* curation independent of prefab containers of nation or religion. The thirst for disseminated knowledge in a technologically social age is akin to the Renaissance explosion post-Gutenberg.

5. Invisibility. Participants can *away* themselves or disappear at any point. Although it can be over-ingested, social media offers an accessible voice for the otherwise shy, potentially facilitating precise and holistic critique leading to reformations that more pressurized environments may have precluded.

6. Personas. A big perk of creative collaboratives is their ability to design and project a unified face, controlling identity, appearance, allegiance, voice, interests, et al. Dictating peer perception boosts self-esteem, allowing members to ensure that their contributions are represented.

7. Watchmen. Technology comes equipped with a posse. People are keen to keep abreast of the social components of their own work and the involvement of their peers, jumping on the opportunity to fix mistakes when representations are inaccurate. When given the self-governing responsibility of checks and balances, collaborators are nosy kindred with a mission.

8. Copyleft. Tech collaborators have a commitment to uphold Creative Commons licensing and other tools for building shared knowledge, allowing culture to build upon its past, as documented through works like Brett Gaylor's *RIP: A Remix Manifesto.* They understand business models are changing and advocate digital development and open-source technology.

"[A]nyone with an Internet connection can make their creations available to the public, unmediated by the old gatekeepers of mass media. The result has been an unprecedented outpouring of creative works. ...As the best of those works are now making their way into the broader cultural landscape, they're breaking mass media's stranglehold on the ownership of meaningful content. ...The dirty secret of mass media, though, was—and still is—that a great deal of it belongs to the companies that distribute it, rather than to the people who make it. That's begun to change as the Internet rewrites the rules about who can put creative work into the public sphere as well as who can take it out."—"The Future of Open Source: Collaborative Culture" by Douglas Wolk of *Wired*

9. Organization and Mediation. Tech-acknowledging collaboratives have an inherent benefit due to structural traits that reduce overhead costs and increase interaction; pre-formatted social springboards kick-start collaboration, layer and catalogue communication, promote efficiency, and require low maintenance.

10. Customizable. Group members can personalize their social settings and make their collaborative environment as comfortable and productive as possible. Modifying the user interface, requesting hazelnut coffee, acquiring standing desks or Eames chairs for the workspace—collaborative cultures provide the tools and atmosphere to encourage members to move in and own their work world.

11. Responsibility. Neil Postman's warning that "technology giveth and technology taketh" emphasizes the need to maintain perspective on each facet of technological socialism. Power, as Spiderman repeatedly learns, must be tied to responsibility. Tech collaborators must insist that technology remain a tool to promote meaningful human interaction. Accepting any technology must be weighed, not assumed; and when tech is utilized, it is for group benefit, not individual benefit. Beyond the prevalence of winners and losers, with each tech adaptation, aspects of culture are promoted or diminished.

Interactivity vs. Connectivity

Fact: If you die in Facebook, you die for real.

While technological socialism has spurred the accessibility of information, ease of collaboration, and horizontal structures, it has also convinced Generations Y–Z that connectivity is the same as interactivity. Growing up on mediated human interaction has its consequences, and the fallout manifests in everything from communications in the workplace to downtime spent with friends. Without chat bubbles, faux 3D buttons to tap, punctuation stand-ins for emotions, Sims-synth soap operas, and popularity scorecards, people are forgetting what physical proximity entails. In *The New York Times* article "Out on the Town, Always Online," John Leland records typical, casual conversational patterns as interrupted by tech's intrusions. Leland's stories are the mundane lunacies of modern interaction: friends sit at tables and speak in fragments as phones beep and draw attention to exterior conversational fragments, all with requisite pretend bite-sized laughterlies. In person, conversations resume, scattershot and contentless. Digital information intrudes in spurts, also mostly contentless. Attention—of all parties—wanders. The point of Leland's sampled profiles of New Yorkers is that there is no point; seamless living in the physical and digital worlds simultaneously spreads experiences out to the level of absolute dilution.

Ideally, technology amplifies the reach of physical interaction, looking to simulate, improve, monetize, and replace it when convenient—this is nothing new. Tech tools and companies are dictating culture and aggressively modifying interactions. Not to say that socially acceptable elevator conversations about the weather are particularly meaningful; but the constant sharestream has little human benefit, while greatly benefiting corporate interests, and that is worse. The very technology that enables interdisciplinary work by making information and tools available to young designers, that enables those designers to work with other Makers around the world, also flattens out the value of interaction for anyone not holding the wires.

maybe if you twist a little to the right?

CONNECTIVITY

"Among the famous aphorisms from the troublesome pen of Karl Marx is his remark in *The Poverty of Philosophy* that the 'hand-loom gives you society with the feudal lord, the steam-mill, society with the industrial capitalist.'...Marx understood well that, apart from their economic implications, technologies create the ways in which people perceive reality, and that such ways are the key to understanding diverse forms of social and mental life."-*Technopoly* by Neil Postman

Re: video game, *The Sims*: "It's not a game about managing life (like *SimCity*) or even creating life (like *SimEarth*), it's a game about experiencing life, and experiencing it in the most mundane fashion possible. Whenever unimaginative TV critics tried to explain the subtle, subversive genius of Seinfeld, they always went back to the hack argument that 'It was about nothing.' But that sentiment was always a little wrong. Seinfeld was about nothing but its underlying message was that nothingness still has weight and a mass and a conflict. What seemed so new about Seinfeld was that it didn't need a story to have a plot: Nothing was still something. *The Sims* forces that aesthetic even further: Nothing is everything."-*Sex, Drugs, and Cocoa Puffs* by Chuck Klosterman

A similar flattening happens in other aspects of culture, including humor reduced to quickly collected memes. Remixing becomes less and less collaged, fewer original works rise to prominence, and the interest in anything nuanced that requires dwell-time with the work is shoved to the margins. Instead of prizing the depth of accessible information, online habits prioritize speed and mass. Trends rise and fall instantly without developing depth or growing over time. New Yorkers claim FOMO as a local disease, but it is a global need—anything hugely popular must be accessed, and instantly—the only explanation for phenomenons like Psy's "Gangnam Style." According to *Adbusters* magazine frontman Kalle Lasn, in his book *Culture Jam,* "All of us somehow felt that the next battleground was going to be culture. We all felt somehow that our culture had been stolen from us—by commercial forces, by advertising agencies, by TV broadcasters. It felt like we were no longer singing our songs and telling stories, and generating our culture from the bottom up, but now we were somehow being spoon-fed this commercial culture top down. ...I see a lot of frenetic activity in cyberspace, but a lot of it is like the postmodern hall of mirrors. It's just people sending email messages to each other, hand on the mouse, and you think that you've done something great if you get some big idea here and send an email to your friend, and pass it on, and you think you have made some sort of a big thing for the day. I don't actually see too many really new ideas coming out of cyberspace yet. I see a lot of new ideas still coming out of philosophers, musicians, thinkers, sociologists, a few economists. I think that the big ideas are still coming out in the traditional way, and then they start to reverberate within cyberspace."

The increasing belief that UX/UI programming specialists are more important than design or vision reflects America's relationship with tech. Paul Graham in *Revenge of the Nerds* sums up the trend well: "The pointy-haired boss miraculously combines two qualities that are common by themselves, but rarely seen together: (a) he knows nothing whatsoever about technology, and (b) he has very strong opinions about it. Suppose, for example, you need to write a piece of software. The pointy-haired boss has no idea how this software has to work, and can't tell one programming language from another, and yet he knows what language you should write it in. Exactly. He thinks you should write it in Java." Even within design, this techflash manifests in the over-reliance of poorly constructed type moving frenetically in title sequences. Legibility is not making way for communication in a Carson-sense. Even Style, which can carry elements of Message, is not at fault—rather, formal and conceptual factors are made subservient to a belief that anything not MOVING or 3D is cheap, or worse, boring.

Connectivity is a business, and users need to be reminded that they are consumers. Connectivity propagates voracious carnivores of conversation. With sinewy bits of undigestible texting-tissue stuck in between our teeth, can we ever separate the medium from our message? There has been an increasingly inverse relationship between the rise in commercial technological connectivity between any two people groups and the depreciating merit of those interactions, making it a double-edged tool of collaboration.

Call to Action: Socratic Media

As deliberate participants in constructing culture, we personally believe in the potential of social media, but the *social* part seems broken. Social, as it exists now, is actually a dependency on corporate companionship through a superficially built community, and participants need to self-question if their involvement counts as creation. Social, as it exists now, is a pseudonym for Mass Appeal. In the *Wired* article, "The Web is Dead. Long Live the Internet, " Chris Anderson notes, "As much as we love the open, unfettered Web, we're abandoning it for simpler, sleeker services that just work." Instead of settling for corporate curation, we suggest *Socratic media,* where all tech tools and toys are investigated before integrating. Unique dialectic inquiry will achieve the greatest prominence, tools will provide users with more control, while users and financiers will encourage designers to execute specific visions. Imagine if popular interactive technology was designed to make us think and to prompt meaningful debate beyond rants regarding chain-produced superhero casting decisions. What if society prized high intellectual standards, informed curation, truly live discussion, and content-driven interaction? Is there a happy balance of smarts and speed? We believe that it is entirely possible to work toward Kurt Vonnegut's anti-loneliness society while maintaining critical thinking and guided collaborative conversations via custom interactive technology.

SIGNAL TO NOISE
Neil Gaiman and Dave McKean
comic dissecting The End

REVENGE OF THE NERDS
for Paul Graham essay, if he wants it

INTERVENTION
Skyler mediates conversation

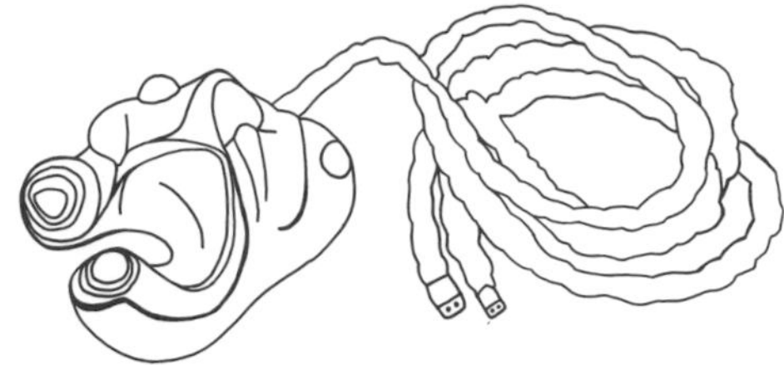

UMBILICAL GAMING
Existenz (1999) from David Cronenberg

ADVENTURE APP
surf's up

UPLOAD SKILLS
characters in Joss Whedon's Dollhouse load abilities from memory

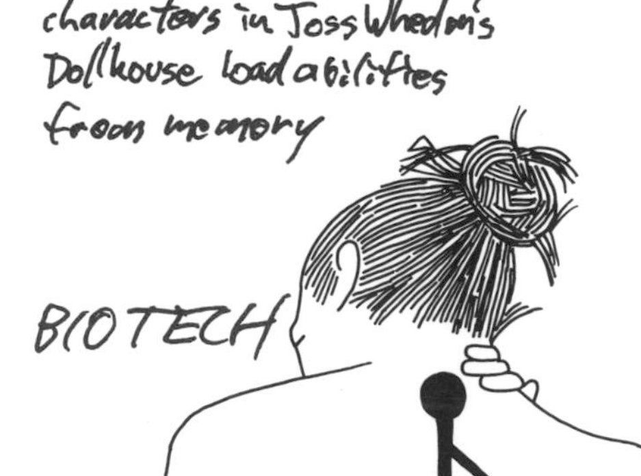

Collaborative Design Technology

Design professors facilitating concept critiques on Facebook, branding designers letting clients in on process work through shared Dropbox folders and Twitter feeds, photographers sharing additional giveaway images on Flikr, animators providing source videos for remixers on Vimeo and YouTube, programmers sharing code bits through Google Docs…

Artists and designers have been hijacking and blending available social technologies for collaboration from the beginning, but collaborative digital design tools are still lacking. For example, in "How to Build Your Own Google Docs (Without Google)" by Cade Metz for *Wired,* the author details how open-source elements can be repurposed for collaborative editing; essentially, tools must be hacked if privacy or control are desired. Makers often resort to abusing a plethora of apps designed for businessmen. These can be lumped into eight categories that facilitate: 1. mind-mapping, 2. feedback, 3. project management, 4. chatting, 5. live text and code editing, 6. live sketching, 7. video and screen-sharing, and 8. application and file sharing. While some of these can be helpful to the collaborative design process, very few address the reality of Creatives' daily needs. A couple examples do seem to have potential, including Firepad and Scoot & Doodle. Firepad is a real-time code editor, and Scoot & Doodle is for live collaborative drawing. Ideally, more inclusive collaborative design tools will arrive soon, especially if they evolve within the preexisting industry standard design programs. Hi, Adobe!

LiveCycle Collaboration Service (LCCS), by Adobe and its partner Influxis, allows businesses to "easily add real-time collaboration and social capabilities to their applications." It sounds fantastic, except it's not. The program does many interactive things; however, none apply to creative outlets and the software mechanics have a rocky history. For example, LiveCycle can help make PDFs into interactive forms, which is not a new thing for designers, nor does it facilitate the creative part of the job. Due to the large programming investment needed to set up LiveCycle, a developer is better off building real-time collaboration into existing applications, such as multi-user whiteboards. But apps like the aforementioned Firepad and Scoot & Doodle, as well as Apple's OS and iOS, which linked chat and sharing tools, can achieve much of this, and the interfaces are intuitive for new users. LiveCycle is developer-focused software. It is not conceived of as a user-friendly collaborative feature integrated into other Adobe design programs.

Another Adobe misnomer is Creative Cloud for Teams. Launched in April 2012, it allows businesses to purchase employee subscriptions to a shared database of applications and company files. Again, the software does not have real-time creative collaboration capabilities. It is like a company locker room. Instead of Adobe Creative *Cloud,* artists and designers need an Adobe Creative *Collaboration.*

James Earl Jones:	Malcom McDowell:
will you go out tonight?	
	probably not buuut, maybe
i'm kinda tired but i also kinda wanna go out	
	yeah! me too!
well, text me if you do	
	k. but i probably won't but i might...

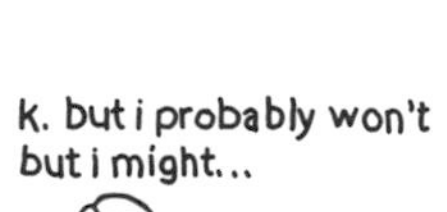

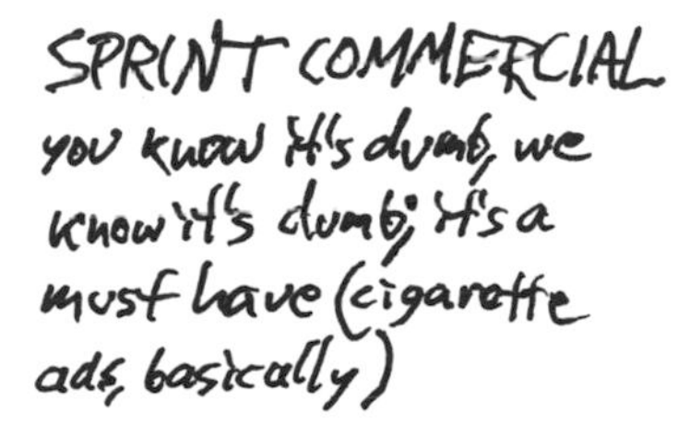

SPRINT COMMERCIAL
you know it's dumb, we know it's dumb, it's a must have (cigarette ads, basically)

LOST IN THE CLOUD
sharing is not collaboration

ENERGY SYNC
waste concept for self-charging

SERVER SPACE

Collaboratives Deploying Diverse Interactive Technologies

HITRECORD

collaborative online production company from actor Joseph Gordon-Levitt and his brother Dan

ISEE

McSweeney's Quarterly

For its first issue in 1998, writer/editor/publisher/activist Dave Eggers sent out a mass email to solicit material rejected by other magazines. Although the curation metric was soon abandoned, it kick-started a submission and subscriber base that included some of the best writers and artists in the States. The response to *McSweeney's* was so high that the endeavor expanded to become a San Francisco publishing house, producing books, magazines, and experimental ephemera with a storefront, t-shirt shop, and eclectic garage sale.

RIP: A Remix Manifesto

Posted in 2008, this open-source documentary film explores in combination: Creative Commons licensing, the Copyleft ideology, and profiles of remix artists, while arguing that the culture of the Information Age must be allowed to build upon the past. The video is downloadable via a "name your price" checkout, and the website encourages users to remix the video and resubmit it to ripremix.com for incorporation within the evolving mash-up documentary.

FEEDBACK from the raste icon set

ETSY we can do it... with a glue gun

HitRecord

Actor Joseph Gordon-Levitt launched this online collaborative production company with his brother Dan in 2005. Writers, musicians, illustrators, video editors, curators, photographers, singers, graphic designers, dancers, performance artists, remixers, and fine artists of all kinds are invited to submit projects. Contributors have access to the whole database of submitted work to remix for their own pieces. Gordon-Levitt highlights the company's objective: "We use the Internet to collaborate with artists from all over the world and we form a community around all of the work that we make together." HitRecord has screened at Sundance Film Festivals, published books, released records, sold t-shirts, printed posters, played live shows, and hosts a half-hour variety TV show on Pivot. HitRecord pays collaborators for their creative contributions.

Linux

In 1991, Linus Torvalds launched a beta version of this open-source operating system that would eventually be used by millions of people worldwide. In the *Wired* article "How Linus Torvalds Became Benevolent Dictator of Planet Linux, the Biggest Collaborative Project in History," Gary Rivlin writes, "His is a world that works only if the best idea wins; he has no

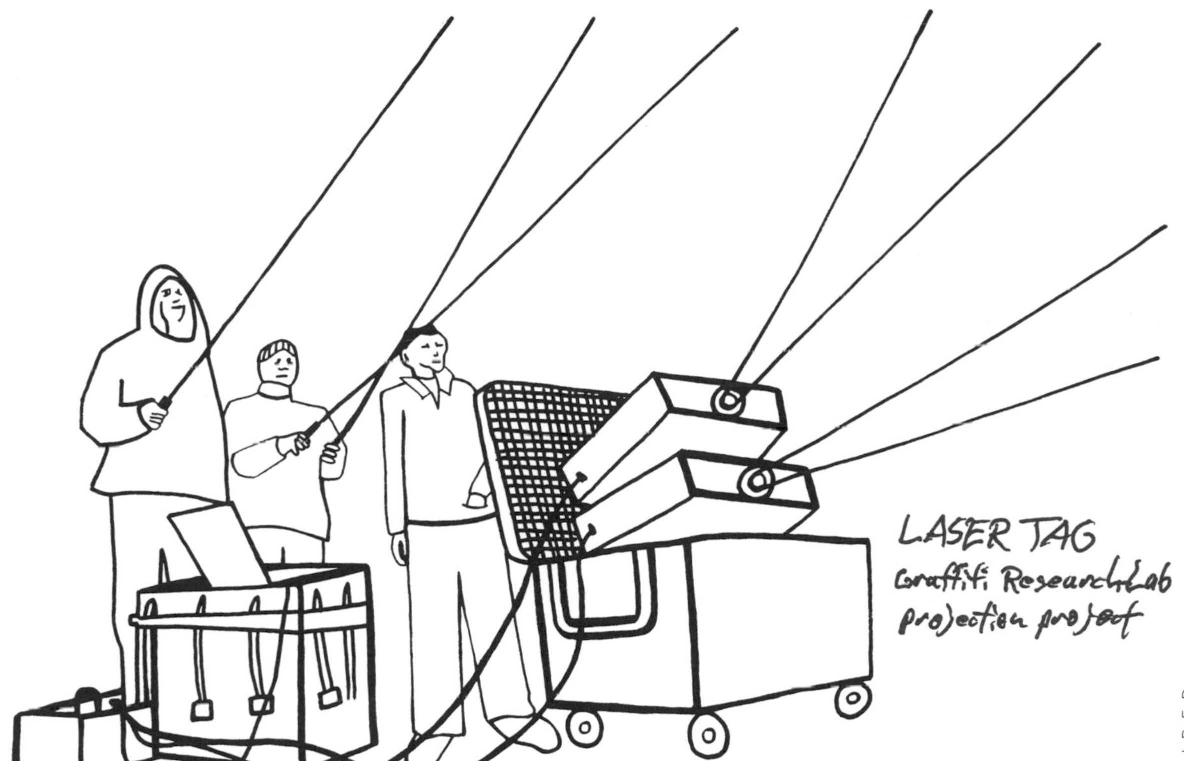

giant marketing budget to compensate for poor technical decisions, no clout in the marketplace to compensate for mediocrity." Torvalds wrote the OS' kernel at 21, released it online for anyone to develop, and Rivlin notes, "Since then, tens of thousands of them have, making Linux perhaps the single largest collaborative project in the planet's history." A huge range of products and projects utilize the tech, and Torvalds delicately manages the loose community that makes it all possible. His "leadership" is an evolving ad hoc collaborative coding committee that communicates primarily via email and annual conferences.

Inside Out Project

Photographer JR started this global photo project in 2011, inviting communities to submit their portraits to insideoutproject.net to be printed—either by live printers on the streets with "inside out photobooths" or by mail—and then installed as massive-format, black-and-white street pastings. The installations are documented, archived, and exhibited online. More than 100,000 posters have been created in more than 108 countries.

WordPress

This free and open-source blogging and website-building tool was first released in 2003; today it ranks as one of the best-loved and most user-friendly code/design tools, deployed across the broadest spectrum of sites. Users of WordPress connect via WordCamp Central (conferences), WordCamp TV (video blog), WordPress Core (blog), IRC (Internet Relay Chat), and email to work together to streamline code, share ideas, fix bugs, and hang out. WordPress is licensed under the General Public License (GPLv2 or later) which provides four core freedoms. Consider this as the WordPress "Bill of Rights:"

+ The freedom to run the program, for any purpose.
+ The freedom to study how the program works, and change it to make it do what you wish.
+ The freedom to redistribute.
+ The freedom to distribute copies of your modified versions to others.

StoryCorps

RECORD YOUR STORY
travelling audio library
collection bus

TAXI: The Global Creative Network (designtaxi.com)

A cutting-edge global arts network, the site launched in 2003, its mission statement to establish a "meeting grounds for creative professionals around the world to connect with each other, stay updated, and showcase their works towards prospects and opportunities. It is a daily-updated international creative media that bridges diverse creative and design disciplines and promotes collaborative interaction to propel unlimited innovations and infinite breakthroughs." TAXI solicits editorial, creative, and inventive material from design businesses and schools via email submissions. The results are highly interdisciplinary, as all perspectives on illustration, product design, photography, and graphic design take turns in the spotlight, while various portfolios and Makers are called to the fore; well-known Creatives and big-budget projects are gridded next to inventive student concept work.

StoryCorps

They come in a bus and set up their recording booth in a city near you and listen. Since 2003, this non-profit endeavor has been collecting America's stories for safekeeping at the Library of Congress' Folklife Center, while broadcasting weekly on NPR's *Morning Edition.* Their mission is to give anyone "the opportunity to record, share, and preserve the stories of our lives." Believing that people are intrinsically interesting and natural storytellers, StoryCorps reminds us of the value of listening to one another in a culture that is often too busy and distracted to do so.

Conclusion

If we've learned anything over the past 20 years since *Geocities,* the first social media site, it's that the Internet is still a very young collaborative technology. As such, interactive tools for design have yet to be fully realized. Our culture is wired and waiting; the hacker community and adventurous entrepreneurs are working hard to combine and improve tech capabilities that maintain the democracy of interactive outlets. In the meantime, it is important to promote a Copyleft mentality and encourage frustrated designers to come together to invent collaborative technology that responsibly equips for intuitive creative interactivion.

"You have to design for them. People are not robots, and it's rarely a one-to-one relationship between question and answer.…So how can brands build and maintain strong relationships with the customers who matter most? It starts with overcoming a problem that plagues most businesses: a lack of intimacy."-"XBox One, Netflix, Charles Schwab: Why Consumer Collaboration is Key in Business" by Dmitry Dragilev of *Wired*

UN

JESSICA HISCHE
+ RUSS MASCHMEYER
y/n?
HOME STUDIO

Designer Couples- Co-Artship

"It is harder for a woman to get equal credit for her work in a husband-and-wife team. Oftentimes people just assume-particularly in academia-that the male partner runs the office while the female partner teaches."-Eric Höweler of Höweler and Yoon Architecture

Kids, let me see your hands. Jill, is that toner under your nails? I don't even want to know. Go scrub with the Gojo, then let's eat.

Did you set the table, Jesse? There's a grid right on the tablecloth, but these bowls are all over the place.

Honey, you should see the run of invitations we got back from the letterpressers today. They're immaculate. Look, that impression, the paper, the coloring, it's perfect for this new set of wood type they just acquired.

That's lovely, but how did you get away with running your type backwards and no title on the front?

Well, it's this little museum....How was your day? And pass the carrots, please, Spencer.

I had to be in early to meet up with our visiting designer over hummus and bagels, but we were late because someone managed to pick the one pair of red sneakers in the universe that did not weigh enough to pair with their toile-patterned skirt, ahem, Jill. I swear that hell is having a closet full of black and none of it matches.

Dad, can I go play with the laser cutter after dinner? I want to make a nameplate for my bookbag.

Fine, but don't let your sister trim the dog's nails with it this time, Spencer. Hey babe, where are we at with the numerals?

I did a bunch of studies with the paint color of the doors and everything I could find in the hardware stores downtown. We're going to have to have numbers made. Nothing matches the forms of the trim out front. Sketches and shots are pinned in the studio downstairs, let me know what you think.

OK-on my way...

Oh, and Sweetie, leave the furniture alone tonight, would you? It's fine where it is, and you scraped the floor last week when you were moving it.

I love our design family, even you, Jesse. It's just so damn convenient.

Sociometry

Collaboration is an attractive ideal. But more specifically, colleagues in various scenarios—office peers, classmates, bar regulars—are often attracted to each other. Usually it starts due to propinquity, as a result of a shared social situation. When grade school teachers assign seats according to alphabetical order, chances are Voynovich and Zimmerman will talk to each other over the course of the year, suddenly creating the possibility of friendship. People respond to the demands of their environment, and that shared space, objectives, and revealed overlap can appease tensions that normally block the development of interpersonal relationships.

Says Graphic Designer Rudy VanderLans, "Zuzana and I met at the University of California at Berkeley in 1981. Zuzana was studying architecture as an undergraduate, and I was in the graduate program studying photography. Both of these departments were within the School of Environmental Design, and all were housed in the same building. So we often ran into each other in the hallways, and it didn't take long before we realized that we were two odd ducks, because our real interest was graphic design."

In *Our Social World,* Donelson Forsyth, social psychologist and the Endowed Chair at the University of Richmond, discusses how people placed into social situations are constantly bombarded by *jolts* of self-awareness, reminders of their presence amongst company. Specifically, the presentness of people functions as an alert. A social situation is one of uncertainty in which people are primed to be responsive and ready. Once aroused, senses are heightened and people tend to improve their performance. This involuntary arousal reaction tends to push people to work longer and better as alertness becomes cyclical.

As shared projects gain momentum, social awkwardness dissolves in the frenzy of production. Collaboratives contextualize work as a game instead of a chore, helped along no doubt by any attraction motivators. Collaborators become

inspired by each other, by their ideas and effort, and the work becomes as important as earning respect. Attraction is most effective in this approval process.

The argument goes that if Tom tells Anna her work is fantastic, but Dan tells Anna she can improve her composition, Anna will instinctively feel motivated by Dan, ignoring the already vanquished Tom in favor of the more interesting pursuit, rising to the perceived challenge. After four revisions, Dan tells Anna her work is fantastic, and Anna feels more inclined to like Dan than Tom even though both Dan and Tom like Anna's composition. Dan will also feel closer to Anna for having found an open and honest working relationship. Aw.

Donelson Forsyth writes, "Attraction obeys the *reciprocity principle:* we like people who like us." He defines the *similarity-attraction effect* as, "We like people who endorse values and attitudes that are similar to our own. ...Similarity even promotes romantic attraction...when other people agree with us, they confirm the accuracy of our beliefs." Collaboration often reinforces the participants' charged energy and even physical attraction toward other group members. Working with a colleague who's equally invested on a project, contributes to emotional bonding. Design relationships often follow this path: shared challenges lead to affiliation to empathy to compassion to affection. Creative inspiration grounded by mutual love leads to more compassionate, productive work. Sociologists define *ultimate relationships* as a combination of friendship, passion, and committed work.

Opposites do not attract, argues John von Neumann in a 1928 study conducted by multiple social psychologists regarding various social games and values. Attraction that leads to love is based on similarity, not dissimilarity. Decisions are based on minimizing loss or maximizing gain. The *minimax principle* implies that people are attracted to those who offer the most reward with the least costs. This accounts for the number of design couples. Combining professional savvy, interests, and support in a context that recognizes the demands of a design career creates a strong relationship.

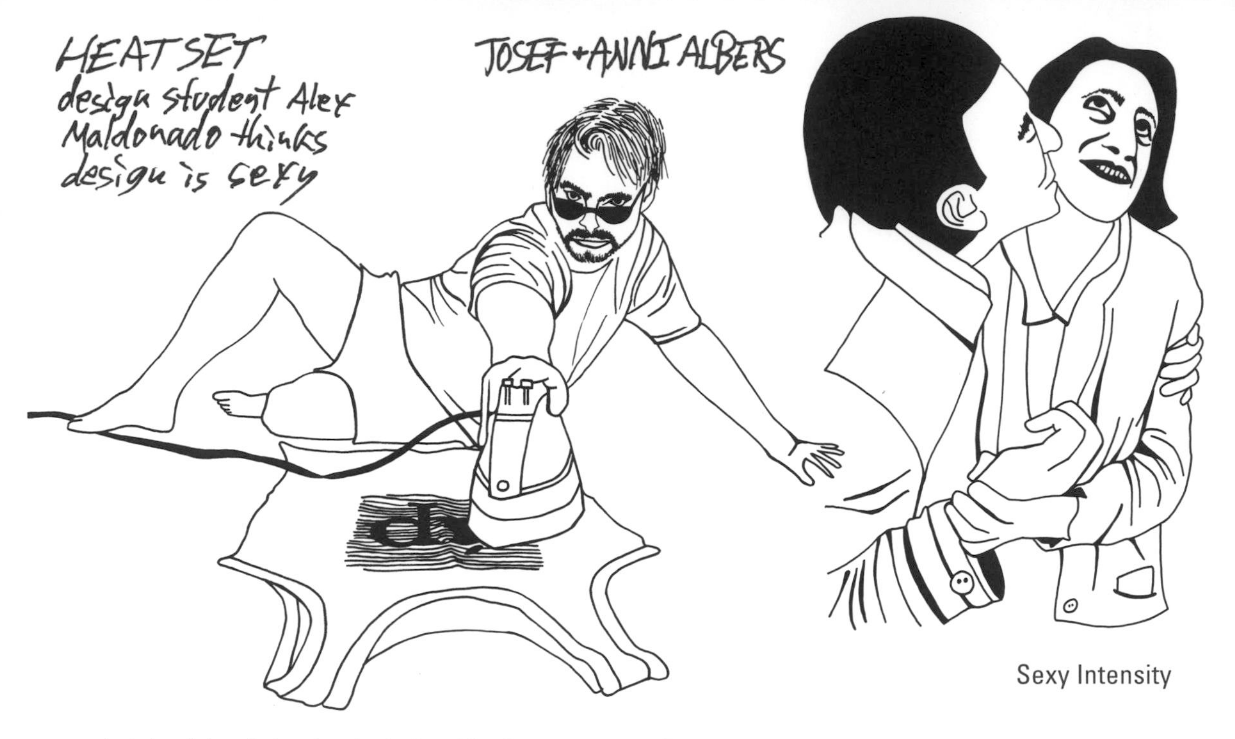

Sexy Intensity

In their collaboratively written book, *Organizing Genius,* Leadership Studies pioneers Warren Bennis and Patricia Ward Biederman identify trends and key qualities of some of America's most innovative collaboratives, such as Disney, Apple and PARC, Clinton's campaign war room, The Skunk Works, Black Mountain, and The Manhattan Project. They found that the groups were marked by an overt sense of "fun." Stress relief meant everything from conference room ping-pong to slugging colleagues with raw eggs to chair racing to late night pub crawls to erotic orgies. Grueling work makes offline commiseration and the comfort of company a source of strength. Oftentimes, this becomes essential. Josef and Anni Albers met while Josef was a professor at the Bauhaus during the turmoil of WWI, and he married Anni, a student, in 1925.

The blend of silly and serious, precise and juvenile, allow groups to function through intense projects. Sexual release and romantic distraction provide tools for brains to bond and expand outside of the narrowness of the group's focus. Bennis and Biederman recount the infamous Snow White Orgy at Disney: "On at least one now-notorious occasion, the high spirits got embarrassingly out of hand. To thank the staff for their heroic effort in finishing *Snow White* on time, Disney invited everyone to an all-expense-paid weekend at a hotel near Palm Springs. The champagne flowed freely, someone jumped naked into the swimming pool, and soon the men who had drawn the Seven Dwarfs were cavorting with the nubile young women who had inked and painted them." Walt and his wife were horrified. But, as Bennis and Biederman point out, the party is an air release, an escape valve, for all the energy and interests pushed to the side for so long during the project's run, and it has a parallel in the famous pranks, fun environments, and game-playing in today's tech industry.

All this pressure-cooking can fracture relationships outside the artificial bubble of a project. Girl/boy-friends, spouses, and children are often neglected, replaced by internal interactions. At Black Mountain, John Andrew Rice and Charles Olson, two of the three leaders at the school, had affairs with students. Olson left his wife and daughter for student Betty Kaiser, and they lived together on campus with their infant son. Says Bennis, "People trying to change the world need to be isolated from it, free from its distractions, but still able to tap its resources. ...Great Groups create a culture of their own." Highly invested collaborators often pay a price for their membership. Collaborative bonds are like potent drugs. After a project is over and the group dissolves, postpartum depression can make everyone and everything else seem utterly boring. Thus many collaborators try to keep that addictive energy alive through new projects, online relationships, and courtship.

Internal Worlds of Art Couples

Because art and design careers are often very time- and labor-intensive—requiring experimentation and unexpected solutions, in many cases without requisite compensation—a colleague with a similar background brings an innate understanding of priorities and values that can seem foreign to outsiders. Even artistic spouses and family can fade from view as Makers create internal worlds, while the line between Work and Personal dissipates. If art history is any indication, this contributes to affairs, ill-advised romances, and divorce.

Photographer Edward Weston is a telling example. A romantic and a womanizer, his charm-laced letters and flattering photographs captured the hearts of many women who served as models, studio technicians, muses, and lovers. In 1908, Edward married Flora May Chandler, his sister's best friend. Flora had inherited some money from her father when he died, and she gave Edward an allowance that financed his photography as a full-time pursuit. Flora raised their four sons together over the course of a 29-year marriage, while Edward lived everywhere except home, hard at work away from his family, moving throughout the southwestern United States for 16 years. While his wife invested in his family, Edward moved between at least five notable colleague-lovers. He fell particularly hard for Margrethe Mather, a more experimental image-maker than Edward, influencing his shift from soft-focus pictorialism to sharp-focus modern work. Their relationship became a partnership, and they began co-signing prints. Tina Modotti, a muse and model, became an accomplished documentary photographer working with Weston in Mexico during the '20s revolution. As their creative paths split, Edward's new model/muse Sonya Noskowiak also engaged in a reciprocal art/lover relationship. At 40 years old, he left her for Charis Wilson, age 19, another of his models. They fell in love, and Wilson became a documentarian of his work and travels. In his book *Edward Weston: A Photographer's Love of Life,* Alexander Lee Nyerges documents the exceptionally inspirational role these women played in Weston's work, indicating that the photographer recognized their importance as well. Weston's infatuations with women and photography and are intrinsically bound together.

That romantic inspiration can influence the work to feel less like *work,* as opposed to feeling like a Job, is an inescapable infringement that follows designers home. Like other forms of collaboration, couples often possess an unconscious desire to impress each other. Reflecting Weston, English painter Ben Nicholson married three influential female artists over the course of his career: Winifred Roberts (1920–1938), Barbara Hepworth (1938–1951), and Felicitas Vogler (1957–1977), distilling specific inspiration and motivation from each relationship. For such Makers, there is no toggle switch between personal and professional. Nicholson's career was a seamless transition-merger of a love for working with his models and a love of his camera.

Design brought career-driven superstars Aline Bernstein and Eero Saarinen together. From 1948 to 1953, Bernstein was the associate art editor and a popular art and architecture critic for *The New York Times.* On a business trip in January 1953, she traveled to Bloomfield Hills, Michigan, to interview Eero Saarinen about designing the General Motors Technical Center, which was awarded the most outstanding architectural project of its era by the American Institute of Architects in 1986. According to the *Aline and Eero Saarinen Papers* in the Archives of American Art (AAA) at the Smithsonian, on their first day together, Aline and Eero had dinner, then hurriedly made love in a dark coat room on the Cranbrook campus. Their shared passion of architecture was an aphrodisiac and perhaps a prerequisite of love at first sight. Each were married with two children. Eero's spouse was a wealthy sculptor, but Eero wanted a partner in life and work; Aline's husband was also an outsider to the shared world promised by Eero. They each divorced their spouses in 1951, married, moved to Detroit, and became high-profile collaborators while also maintaining separate pursuits.

A similar situation happened with Eero's closest friend, architect Charles Eames. After twelve years of marriage, Charles left his wife, Catherine Woermann, and daughter, Lucia Jenkins, to marry his Cranbrook colleague Ray Kaiser in 1941. They moved to Los Angeles, where they lived and worked together closely until Charles died in 1978. Their domestic and work lifestyles were so completely integrated that they regularly wore matching or

complementary clothes. In the "Lifelong Collaboration" section of *Charles and Ray Eames: Designers of the Twentieth Century,* author Pat Kirkham writes that: "The sense of coupling, bonding, 'togetherness'—call it what you will—between Charles and Ray was extremely strong and frequently remarked upon. They were often photographed with Charles' arm around Ray, smiling at each other with hands touching or both touching the same object."

Charles and Ray's relationship possessed a charisma that inspired their colleagues at the Eames Office, a direct offshoot of their mutual enthusiasm for design and architecture. Their relationship has many contemporary parallels, including architecture couples Elizabeth Diller and Ricardo Scofidio, Denise Scott Brown and Robert Venturi, Dan Wood and Amale Andraos, Billie Tsien and Tod Williams, Mike Simonian and Maaike Evers, and J. Meejin Yoon and Eric Höweler.

The Georgia O'Keeffe/Alfred Stieglitz collaboration is a unique love story. Like Weston, Stieglitz married for financial security in order to pursue his interests in photography. But since childhood, Alfred was jealous of his twin brothers, Julius and Leopold, who had a very close relationship, and wished for a partner of his own. Alfred resented his wife, who was nothing like him, leaving him missing a sense of collaboration. He abandoned his wife and daughter without regret to work full time at the Camera Club pursuing photography.

In 1916, Alfred exhibited drawings by Georgia O'Keeffe at his 291 Gallery. This kick-started a collaborative affair, in which the two artists corresponded long distance for two years. In 1918, Georgia accepted Alfred's invitation to move from Texas to New York, and they took an apartment together. Four years later, Alfred divorced his wife and married his lover. O'Keeffe inspired Stieglitz's work: he photographed her hands and full nude portraits regularly. Meanwhile, O'Keeffe's paintings shifted from abstract organic form to the geometry of New York skyscrapers. In 1929, Georgia's need to find new inspiration beyond New York landed her on a trip to New Mexico with her friend Rebecca Strand. Together, they started studios and backpacked the countryside. Georgia spent a good portion of every year until 1946 traveling the Southwest, and in 1949, a few years after Alfred's death, she moved to Abiquiú, New Mexico.

"Believe in your partner. It's one of the most amazing feelings to be in a relationship with someone who supports your creative endeavors completely."
-photographer Anna Wolf, wife of designer Mike Perry

GILBERT + GEORGE
photo installation team

What made their collaboration unique was that the majority of their time was spent away from one another. Most of their collaborative inspiration happened via mail. In *O'Keeffe and Stieglitz: Intimacy at a Distance,* Deborah Solomon writes how "O'Keeffe's affectionate stream of letters revived [Stieglitz]. What did he see in her? A gifted artist, a daring exponent of abstract painting, but also a fantasy of innocence that aroused in him a sense of excited paternalism." The long distance relationship made the pent-up desire of their collaborative courtship and correspondence all the more emphatic, but it also provided space for creative energies to explore and explode. Mail, as a medium, forced them to design the content of their interactions, itself a creative act.

Likewise, the relationship between Fluxus artist Yoko Ono and musician John Lennon began at a preview of her 1966 London exhibition, continuing for three years via telephone correspondence and strategically planned meetings, while the distance channeled the desire into creative output. John left his wife, Cynthia, in 1969 and married Yoko. The impact the couple had on each other's artistic career can be seen in John's writing for The Beatles and the performance art demonstrations the couple orchestrated together, including the memorable *Bed-In for Peace,* staged on their honeymoon in protest of the Vietnam War.

"The best thing about working with your partner is that it's very convenient." -Murray Lightburn, husband of Natalia Yanchak, band members of The Dears

Collaborative Convenience

While some Creatives draw support and inspiration from a diverse personal life separate from their jobs, many designer couples comment on the benefits of a streamlined work/life blend. For art/design couples, finding ways to prevent a sense of competition is key, especially in domestic arenas that could translate as a design issue: cooking, furniture, decorating, home improvement, whether the kids should read ugly textbooks.

Some potential tensions come from outside the family. Employers can be skittish of investing in couples, concerned that home issues will trickle into work life. Conversely, an enterprising couple might exert more unanimous influence on the direction of a company.

Design couples often sidestep these issues by being self-employed, either by starting their own studio or freelancing; few Creatives are employed together as Makers or have the same day-job employer. When they do work for someone else, couples often create rhythms where they are not both present at the same time, collaborating through other outlets. Collaborative couples also are all-in, running their career investment without a backup drive. Depending upon the studio/home arrangement, they risk the tendency of introducing variables from their creative outlets that negatively impact the relationship, or vice versa.

Wolf/Perry address this issue: Anna Wolf, "It's easy to get comfortable. …I have a tendency to be anti-social at times, since so much of what I do is meeting and working with new people constantly. So when it comes time for my personal life, sometimes I just want to check out and hide." Her husband and graphic designer, Mike Perry, follows up, "There are moments when we hang out all day, we get home, we're not excited to see each other the way I think some couples feel when they don't see each other all day. That said, unless we are working on a project together, we do our separate thing. We then usually have a day's recap over dinner."

Designer couples find inventive means to take advantage of collaborating at home without losing their edge. Armin Vit and Bryony Gomez-Palacio, authors of *Graphic Design, Referenced,* both work and live together at UnderConsideration, their 2,000-square-foot home and studio headquarters located in Austin, Texas. The master bedroom was converted into the studio, and their two daughters share the third bedroom. The convenience of their combined relationship plays a big role. Being able to do a load of laundry while working with family in the studio, to being half-time parents, and deducting expenses when it's time for filing taxes are all inherent facets of their setup. This lifestyle allows them to enjoy the process of efficient accomplishment. Say Vit and Gomez-Palacio in an interview on "Design Couples" by Caitlin Dover for *Print Magazine,* "Since we are both graphic designers it simply allowed us to focus on this profession with more determination and ambition, especially when we decided in 2007 to quit our respective jobs and devote our time and energy to our clients and ventures." Says Gomez-Palacio, "[W]e can be completely honest with each other while voicing our opinions on each other's work, a partnership that allows us to push and challenge each other in ways that traditional colleagues can rarely accomplish."

Master Printer Erika Greenberg-Schneider and Sculptor Dominique Labauvie share a similar mentality. Located in Tampa, Florida, their headquarters, *Bleu Acier,* functions as a printshop, metal shop, gallery, and home. Their lifestyle seamlessly supports multiple endeavors and collaborative capabilities while welcoming friends, family, and clients for dinners and long discussions in their kitchen/library.

Emigre partner Rudy VanderLans on his collaboration with his wife, Zuzana Licko, says, "Our working and home lives are fully integrated. There's no time clock to punch after we climb the two flights of stairs in the morning to the top floor of our house where our offices are. ...The work we create, our photos and ceramics, are all over our house." For the Emigre couple, work happens over meals, while watching TV, and during domestic chores; there is no attempt to get away from it for home time.

Women in Collaborative Couples

Historically, men have received proportionally more credit for work that has been accomplished as a couple. However, there have been a few notable instances when men have spoken to the public's misunderstanding of a couple's collaboration, or even promoted their wife's contributions above their own. Such is the case with Charles and Ray Eames. Because of her extraordinary sense of color and form, Ray's contribution to the Eames partnership was marginalized by the public as "decoration," "prettiness," and "beautifying." Furthermore Ray battled entrenched public preconceptions; less important *domestic concerns* were expected roles best left to women. This upset Charles, who constantly pushed the Eames' work as collaborative. According to Pat Kirkham in *Charles and Ray Eames: Designers of the Twentieth Century:* "It was [Charles] who claimed 'anything I can do, she can do better,' opened a major presentation on design with 'Ray can really do it,' and insisted 'she is equally responsible with me for everything.' "

There is often no clear division of labor in a collaborative, making it difficult to fight established notions of gender roles by pointing to concrete examples. Billie Tsien, architect and wife of Tod Williams, had a rough start too. In *Couples Who Build More Than Relationships,* Robin Pobrebin explores how Tsien was assumed to function more in support of Williams and a parallel assumption that the latter was unqualified to develop patterns and work with textiles despite an obvious capability.

Denise Scott Brown, the acclaimed partner of Robert Venturi, said in an interview by Vladimir Paperny, "When you are the wife as well as the partner, people typecast you. You are the handmaiden. Your husband is the design genius, and you're allowed to be the preservationist or the planner—something less—and the notion that creativity can reside in two minds is impossible." These perceptions regarding intelligence, invention, and aesthetics seem particularly out of place in architecture, a field dominated by expensive and complex long-term projects. With so many female partners within architecture collaboratives, gender-based assumptions are not only clearly false but buck the essential values and benefits of working collectively. Finally, regarding aesthetics, American architecture's favorite son, Frank Lloyd Wright, was well known as a designer of patterns in addition to Fallingwater.

Film's great example of overshadowed female collaborators is Alma Reville, wife to Alfred Hitchcock, whose film-editing career preceded her husband's takeover of the industry. In fact, Alfred held off on coring Alma until he climbed to a higher-ranking position as Assistant Director for *Woman to Woman* in 1923. According to their daughter, Patricia, he interviewed Alma for an editor position on the film. After Alfred finished talking, Alma told Alfred that the salary offer was inadequate and politely left the room. He ran out after her and rapidly renegotiated. This was the first instance of the rising legend's respect for the shrewd editing skills of his soon-to-be wife. She promptly became the only person whose opinion mattered to him, and her attention to detail revealed details invisible to others. For example, amidst the crowd's cheers during a screening of *Psycho,* Alma observed that Janet Leigh's throat moved slightly after her fatal stabbing, and she held up distribution until the scene was fixed. According to Stephen Rebello, author of *Alfred Hitchcock and the Making of Psycho,* "She really had a major impact on the film, by just persuading Hitch to back off from his own ego and listen to the idea of somebody else."

Aino and Alvar Aalto were a husband-wife team who collaborated on furniture and product designs. They met when she started working for his architecture firm in 1924. The two fell in love, married, and began their lifelong partnership. The Aaltos are famous for their gorgeous glass designs showcased at the 1939 New York World's Fair, utilized by the Finnish company Iittala, and were later copied by IKEA. But it was Aino who truly launched them to fame, beating her husband in a 1932 design competition. Their series *Aino Aalto* became the first set of designer glass products designed for modern mass production.

Many women maintain collaborative production while also practicing independently. Aline Bernstein, before marrying Eero Saarinen, was the Associate Art Editor and Critic for *The New York Times.* When she moved to Bloomfield Hills to collaborate with her husband in 1954, she continued her journalism career. While maintaining Head of Information Service at Eero Saarinen & Associates, pitching Eero's projects to the publications she had previously worked with, she also published *The Proud Possessors* in 1958, a best selling book on art collectors that was funded by a Guggenheim fellowship she had won. After Eero's death in 1961, she became the first female NBC News correspondent and the network Bureau Chief to Paris. With three children, she still furthered her career without reducing her sphere of activity.

Alternatively, some women feel that partnering with their husband is a way to progress in a field historically dominated by men. Says Elizabeth Diller, from Diller Scofidio + Renfro, "There were so few of us [women] and it somehow fortified us, established a more acceptable context for us to practice. If I had been on my own, it probably would have been tougher." Similarly, Amale Andraos confesses that it took time for her to build confidence as a female architect, and that working with Dan Wood and Rem Koolhaas helped her to develop professionally. Amale married Dan, and today the couple are so intellectually in sync, that colleagues nicknamed them "Danamale."

Choosing to collaborate with a partner in ways that are not tied to compensation or co-owning a business allows women to become independently accomplished, without deliberately taking on gender stereotypes when collaborating with their husbands on exterior projects. When Ellen Lupton and Abbott Miller first started working together outside of Cooper Union, they began a collective called Design Writing Research, a collaborative outlet for fun freelance projects and critical design writing, while they both kept steady design jobs. Ellen is the Director of the MFA Graphic Design Program at MICA in Baltimore and Curator for Cooper Hewitt National Design Museum in New York, while Abbott is a partner at Pentagram in New York. A converted bowling alley in their home is now a shared office and library. Says Abbott and Ellen, "Our home office is Ellen's primary base, but Abbott is able to work here when he is not in New York. We are thus able to see what's going on, but we're not on top of each other."

"We met as freshmen in college at The Cooper Union in New York City. It was the fall of 1981. We had all of our classes together that year. We were both from out of town-Ellen from Baltimore, Abbott from northwest Indiana, outside of Chicago. A group of us 'new New Yorkers' became friends together. Freshman year, neither of us knew yet if we wanted to be designers. Ellen was into painting, and Abbott was drawn to sculpture and film. We had a required basic design class with George Sadek, a volatile Czech emigre who loved grids, Garamond, and word play."
-Ellen Lupton and Abbott Miller

ELLEN LUPTON + ABBOTT MILLER

ALFRED HITCHCOCK + ALMA REVILLE

PAUL SAHRE + EMILY OBERMAN

Stephen Doyle, husband of Gael Towey, former Chief Integration and Creative Officer of Martha Stewart Living Omnimedia, calls it the "EHSS: Ex-Husband Surname Syndrome." He first met Gael when being "summoned" to collaborate on a weddings book and remembers her business card with the scratched out last name.

GAEL TOWEY has a practice and media separate from her husband Stephen Doyle

Graphic Designer Jessica Hische co-manages her studio Title Case and conducts typography workshops all over the globe, while her husband, Interactive and Product Designer Russ Maschmeyer, works for Facebook. In their downtime together, they work collaboratively on personal passion projects, such as their websites Don't Fear the Internet and CSS for Typophiles.

Gael Towey worked as the Chief Executive Officer and Editorial Director at Martha Stewart Living Omnimedia for 20 years. Her husband, Stephen Doyle, heads Doyle Partners, a New York–based studio. Says Stephen: "Dear reader: compare and contrast. That's what we offer each other, and that dynamic has infiltrated our 23 years together. ...Gael's empire and my miniscule studio offer us both such insights into design, innovation, energy, and enthusiasm. We are both exposed to very different worlds of creativity when we look through the other's lens. ... And now Gael's inevitable star has risen. Does that threaten me? Absolutely not—I believe I helped her shine."

For contrast, Graphic Designer and Pentagram partner Paula Scher openly admits that collaborating with her husband, Seymour Chwast, Owner and Designer at Pushpin, does not work. In an interview with Caitlin Dover of *Print Magazine,* Paula remembers the few instances when they tried to collaborate: "When I was an art director at CBS Records I hired Seymour to illustrate some covers and it was OK. He asked me to write for the Pushpin graphic, but that was always terrible." When asked how they approach design-related decisions as a couple, Paula responds, "Seymour does whatever I tell him to do." And Seymour echoes, "I just do everything Paula tells me to do." Some design couples simply do not collaborate, though their work bleeds home in more informal ways.

In *The Year of Magical Thinking,* author Joan Didion lets the reader in on the secret of how she worked with her husband, John Dunne, and developed a mutual respect. Her book is a portrait of a successful, although very unusual, marriage in which the relationship is seen as a project constantly under revision. Didion describes one snowy evening by the fireplace on her birthday, a month before John died, where he reads a passage aloud from her book *A Book of Common Prayer.* He closes the book and says, "'Goddamn. Don't ever tell me again you can't write. That's my birthday present to you.'"

TEAMWORK
men tacitly assumed
to do the Real Work
in collaboratives
MADMEN ONLY
christina Hendricks'
Joan Harris character
represents the limited
roles for women in agencies

ANN + PAUL RAND
children's books

Design is One

MASSIMO + LELLA
VIGNELLI

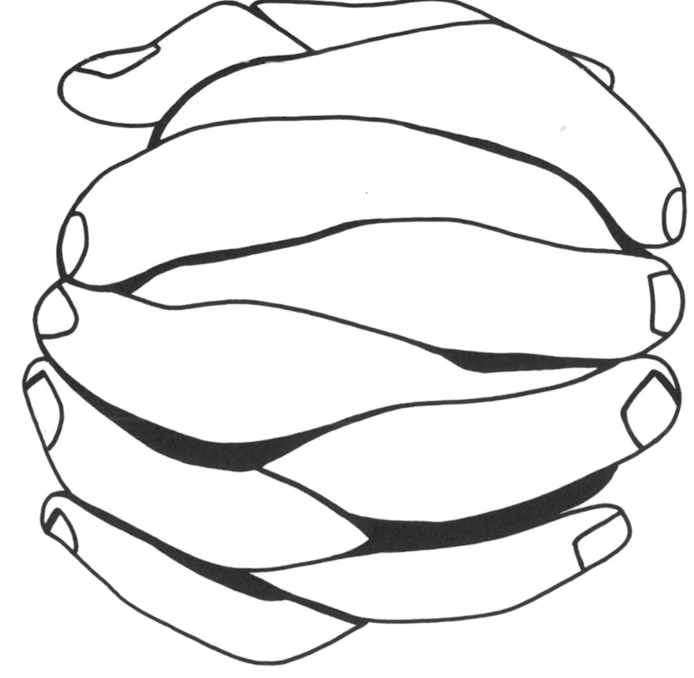

Telepathic Communication

After living and working together, many couples achieve a latent design sensibility that extends beyond complementary contribution and compromise. This seamless authorship is a sort of collision more than simple math. Says Denise Scott Brown, “It's very-very complex. We don't find it easy to tease our ideas apart. Creativity in this office comes from two minds and then from a lot of other minds as well. …But the ideas are shared.” The voice of a collaboration is a unique, birthed entity—an entity unto itself, its own mode of making and thinking, its own message, its own outlook and agenda.

Massimo Vignelli had always claimed that his passion was for 2D graphic design, while his wife, Lella, had a background in 3D, architecture, and form. Massimo is the idealist—inclined towards intuition, possibilities, and what design *could* be. Lella is the realist—focused on feasibility, logic, and what design is. They have worked together since 1957. While skills and styles become easier to integrate through experience, the meshing of ideology —such as the Vignellis allegiance to Modernism— can be a lengthy vetting process, regardless of the collaborators' alignment with things like cultural upbringing and personal beliefs. This is because the collaborative “spirit” is a designed—authored—entity. More than a mission, it's a persona, a designitude developed from what the collaborators need in order to carry out a mission. Say the Vignellis, “When we were young and naïve, we thought we could transform society by providing a better, more designed environment. Naturally, we found that this was not possible. Now, we think more realistically: we see a choice between good design and poor or non-design. Every society gets the design it deserves. It is our duty to develop a professional attitude in raising the standard of design.” Seamless authorship and a distillation of ideology or objectives typically evolves over time within collaborators, as agendas are impacted by shared experience.

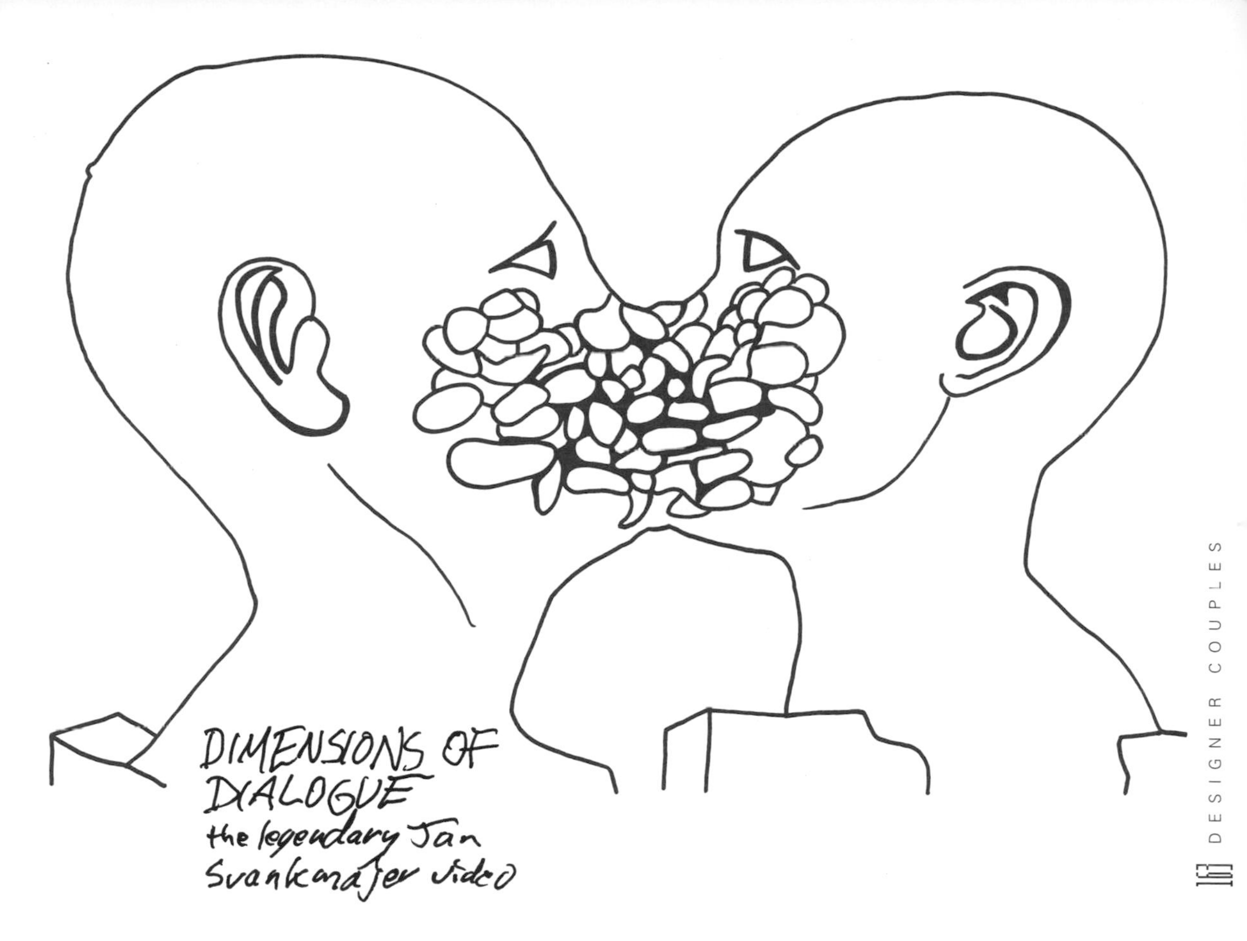

In an interview with LX.TV host Shira Lazar, Christo and Jeanne-Claude discuss their collaborative authorship process: *"Christo:* Of course, this project is very complex, very long. It is not only one person's work, it's really a partnership and collaboration during all these years. …We can be very critical of each other. Actually we're big screamers. All of our filmmakers hear that we're very much arguing all the time. But it has nothing to do with being sick of each other. This is discussing the project. *Jeanne-Claude:* It's a creative screaming. Which is very important for us. For instance, if I shout at him, 'Can't you see that the ropes are too long?' and he answers, 'No, they're too short.' I start thinking on my own, maybe he's right. He's on his side is thinking, 'What if she's right?' And that way, and only that way, can we reach the perfect length."

Like living in shared mind overtime, collaborative couples often develop the ability to communicate intuitively. Pat Kirkham writes in *Charles and Ray Eames: Designers of the Twentieth Century,* "When discussing her own close working relationship with her husband, the architect and designer Alison Smithson used the word 'telepathic' to describe the understanding that develops between two individuals who live and work together. The cumulative effect of interaction between partners is also important. What might have appeared a decision of Charles' could well have been based on ideas exchanged with Ray a day, a week, a month, a year, or several years earlier." Couples that are able to achieve this level of collaborative understanding either have similar personal and educational backgrounds, a great deal of empathy, or have worked together on numerous projects over a prolonged period of time. Or all of these. Because of their ability to anticipate their partner's thoughts and reactions, work can happen more productively, even with minimal interaction.

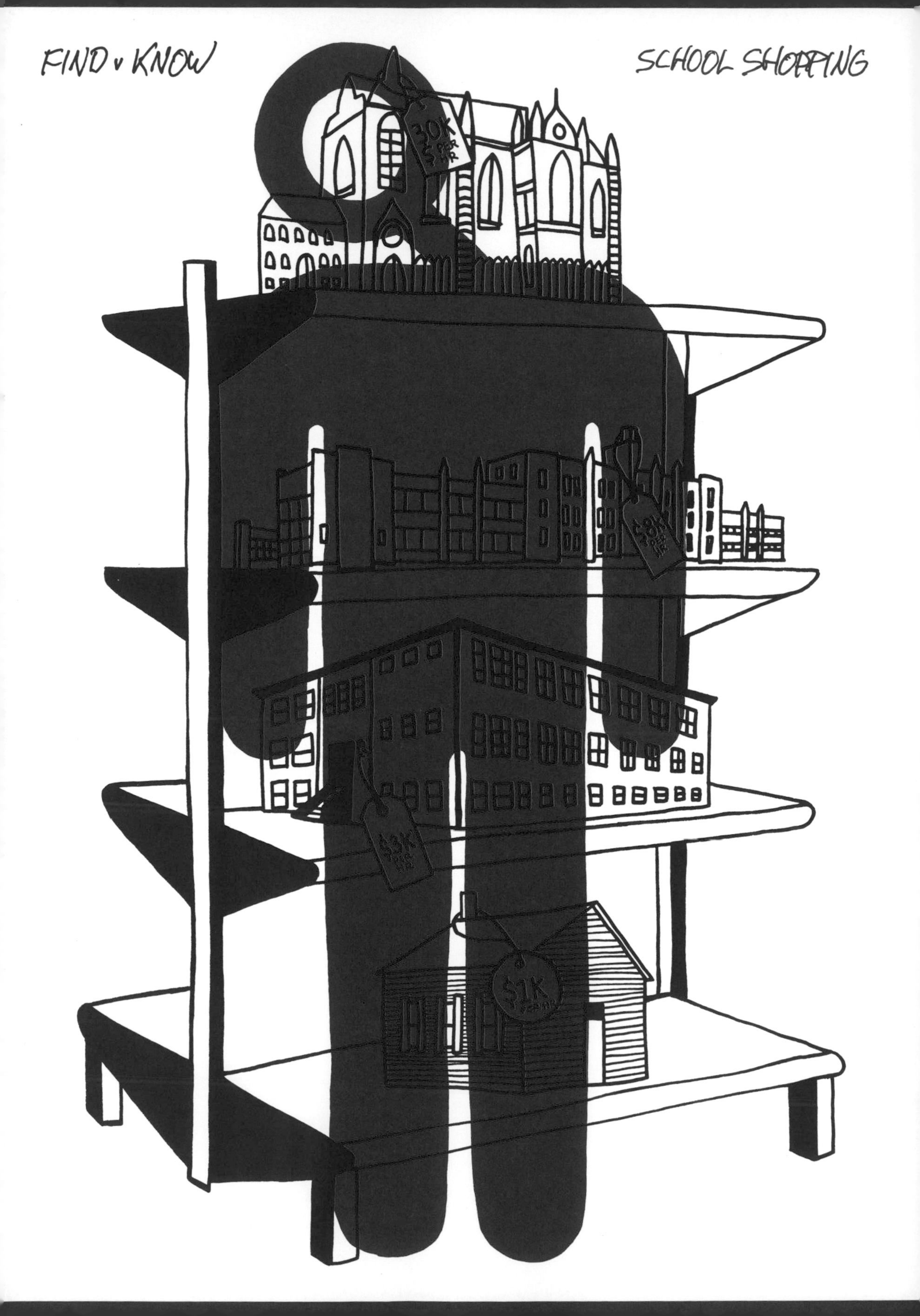
FIND v KNOW
SCHOOL SHOPPING
$30K PER YR
$3K PER YR
$1K

"What Orwell feared were those who would ban books. What Huxley feared was that there would be no reason to ban a book, for there would be no one who wanted to read one. Orwell feared those who would deprive us of information. Huxley feared those who would give us so much that we would be reduced to passivity and egoism."-Neil Postman

In Silicon Valley, Singularity University is experimenting with robot faculty. Isaac Asimov gave us the laws of robotics, Stanley Kubrick eliminated them, and Roger Waters repurposed robot students as ground bologna. Beyond online classes, Silicon Valley predicts the future of schools and professorship as a cyborg hybrid.

This broadcast is live: Cylons are teaching college students. Responding to the demands of overworked instructors who waste time on university service work and clerical duties, neglecting research, Singularity's CEO Rob Nail makes his case in the article "Robot Professors Come With Singularity University's Massive Upgrade" by Ryan Tate: "We really need to have as one of our track chairs an AI [artificial intelligence] faculty member."

Perhaps a hybrid would be better: the subjects of Joss Whedon's show *Dollhouse* upload skills to their brains, providing instant mastery. In his book *The Future of the Mind,* Theoretical Physicist Michio Kaku discusses how "downloading" human consciousness as *information* onto an interactive disc is a real possibility. For example, a student in need of infotainment can dial in a cranky Paul Rand mid-rant to query his thoughts on the updated UPS logo. Kaku states that memory could be "uploaded" into live brains. Psyche will be able to adopt multiple (collaborative?) personas, making the current educational and media industries irrelevant. If the Makers of designer-babies had ambition, they could eliminate the need to learn altogether. Until then…

In the States, educational models are as cyclically fashionable as the Apple logo and as futile as yo-yo diets. Design teachers historically have embraced tech tools, ideally while upholding humanistic inquiry above wired glam. While traditional universities were fudging test scores to boost national ranking, democratic weekend conferences teaching Python, PHP, and digital painting, bypassed state legislatures to reach frustrated artists/students directly. Likewise, free online courses and DIY tutorials became cornerstones in the growing community of open-source Makers. Established schools initially ignored, then absorbed these upstarts.

An historical reliance on reputation, program diversity, and accreditation has proven inadequate, or too expensive, for the lower income Thirsty Intellectuals. The undergraduate Design School is an established tradition, and graduate programs are pushing the standard credentials higher still, as explored by Project Projects' Rob Giampietro in *School Days.* But the inevitability of academia as a necessary industry stepping stone is unwinding, the first reversal of degree escalation since the trade first became an Associate. Schools no longer monopolize Tools Training, so educators have been able to shift from technical craftsmanship to formal and conceptual experimentation. Industry has always prized trendy code wizardry, but as small studios emerge with an emphasis on content, voice, and interdisciplinary collaboration, design education must scramble to adjust the narrow portfolio models currently taught.

1. Tech-ing Effectively

"In the future, social systems will not be adjusted to suit the needs of human beings. Instead, human beings will be adjusted to suit the needs of the system."—*The Unabomber Manifesto: Industrial Society and Its Future* by Ted Kaczynski

$yndicated $chooling

USA TODAY should have died in the '90s. With anchor-adorbs like Tom Brokaw, Barbara Walters, and David Letterman to personally narrate the news at Americans at 30fps faster than a still image, ink offsetting onto breakfast-laden hands seems unjustifiable. In a spot of crisis management, *USA TODAY* responded obviously with color imagery. Then, as a present to itself for its 30th anniversary in 2012, the paper had a color-circle mid-life-crisis makeover, acceptance leading to confidence in its newly curvy self. FOX News Network had a similar freak-out over the speed of news online, and responded with epic iPads and Jumbotrons for their "News Deck," despite the fact that viewers' screens retain their own size. For contrast, *The New York Times* underwent a 2014 redesign after 150 years, but they sided with white space, Cheltenham (an Old Style serif font designed in 1896 and inspired by Will Morris' Arts and Crafts movement), and immaculate proportions with multiplatform considerations.

Print media is not television. Culture critic Chuck Klosterman in *Sex, Drugs, and Cocoa Puffs* makes the case that newspapers contributed to their own demise by trying to imitate TV. Editors demanded bite-sized everything: tiny paragraphs, no articles that spanned multiple pages, color whenever possible, menus for easily finding what you already know to make it easier to skip what you don't. Treating readers as self-loathers who hated reading kept print from capitalizing on what made it different from TV. We are now beyond newspapers but we keep making the same mistake.

Print and TV are not interactive/customizable/searchable experiences. They (ideally) are passive, but controlled, storytelling. *USA TODAY* competed with the web. *The New York Times* maintains a stable of research writers; pushes design and photography specific to its magazine, site, and newspaper; hires hybrid artist-hackers like Jer Thorp as Data Artists in Residence; and commissions rockstar illustrators like Nicholas Blechman, Post Typography, and Al Hirschfeld. Each *Times* venture has a reason for being, with content and aesthetic decisions that play to medium strengths—it is what it is.

Academia, intent on proving its ability to learn the wrong lesson, pimped campus alternatives: online classes, TED lectures, webinars, and MOOCs (Massive Open Online Courses). Campuses are expensive. The national average of the faculty/administration breakdown per school is 1:1 despite the money squeeze, as administrators rarely cut themselves and boards rarely talk to faculty. Blame the '90s arms race when programs and departments could have stood some Roundup, but contemporary economics have not played out as a "be kind, rewind." For example, while faculty share an intrinsic comprehension of the value of multiplicity of perspectives, Diversity Programs and their requisite management persist untrimmed. Costs are cut by automating tasks, outsourcing everything, and replacing tenured faculty with adjuncts and instructors who have little personal investment in the school. (One un/popular saving technique is locking administrative and athletic salaries to the faculty pay structure.) Students don't seem to mind the business model, content to chill at home in their jammies studying. This is all very inevitable; certainly some quality ideas and techniques exist in online academic tools, and no faculty seem to argue otherwise.

VEND-A-PROF
schools are different from businesses

The root conflict then is when schools succumb to consumerist approaches toward education. But School is not Business—the mission of education has nothing to do with money.

American colleges have a particular ability to reduce all academic decisions to "job training," egged on by out-of-work parents, government officials who do not believe in government, and comfy-couched kids. Television, then the Internet, seemed to eliminate the necessity of physically present teachers, conveniently, as schools faced a money crunch. But it is a fair question: "What do schools have left to justify their price tag?"

CODE OF FASHION
Paul Graham argues against trendy construction

Schools must answer, though they often settle for treating education more like a business/entertainment/luxury/vending machine. Raised in a culture of instantaneous gratification, students expect the convenience of online courses from their physical meetings. Design education can be more than bare-bones workforce preparation in a self-serving manner—education is not deer hunting in a "managed forest" with pre-stocked neon-glow

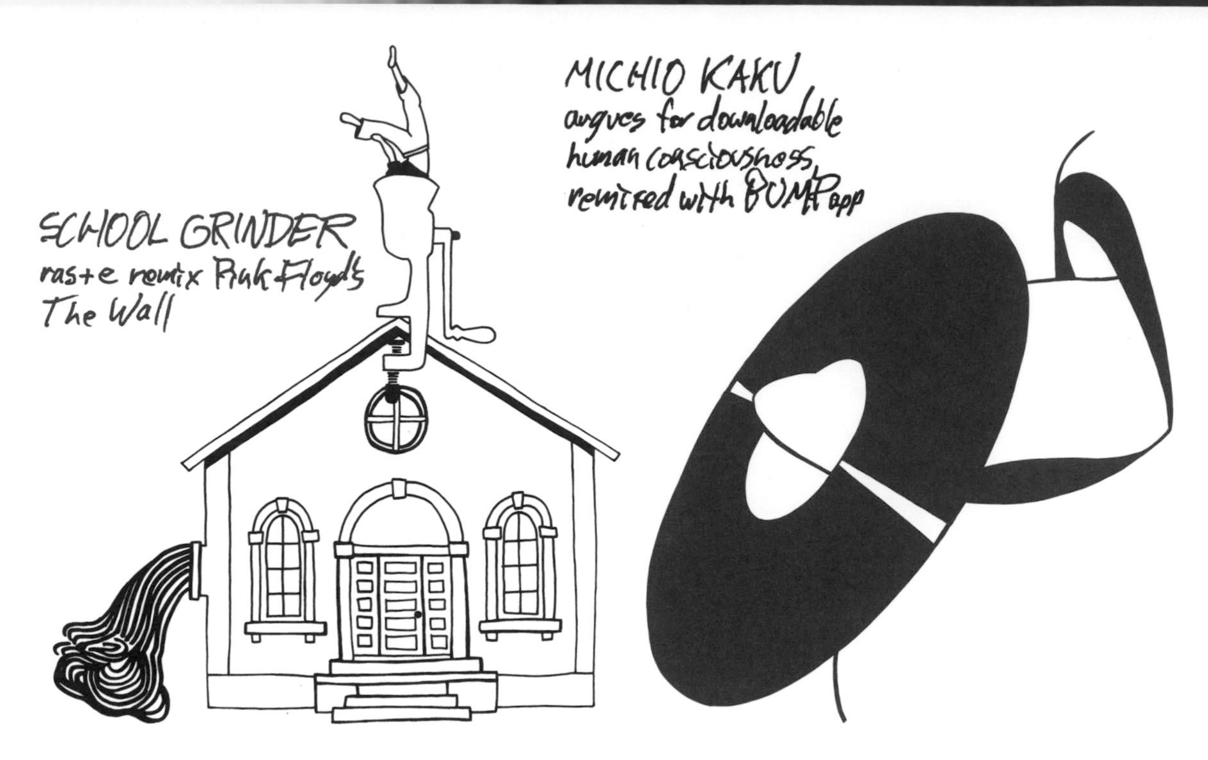

animals, or choosing a Whopper over a Double Stacker at a Burger King drive-through, getting in and out with the most bang for your buck. Education should not be run as a business—there is no relevant culture critic, design writer, or study that disagrees—but in designing design education, the particular advantages of classroom interaction and online tools must be maximized.

In "Can New Technologies Revitalize Old Teaching Methods?" by Pamela Mendels, Lehigh University President Gregory C. Farrington argues that teachers can be slow to add tools: "We've become a bit monopolistic, a bit complacent. We've put too little of our energy into focusing on the challenge of how we create the most effective learning environment at the undergraduate level. We know how we want to teach. We too seldom discuss how do students best learn." Farrington makes the parallel point that certain types of learning primarily occur with in-person interaction. Schools are a natural home for exploring interdisciplinary collaboration, but only if faculty engage with the students, the students engage with each other, and contemporary media are deployed precisely—all of which require time and investment, meaning money, meaning education as education and not education as business.

In his article, "Instruction for Masses Knocks Down Campus Walls," Tamar Lewin writes, "The current, more technically focused MOOCs are highly automated, with computer-graded assignments and exams. ...The Stanford MOOCs, for example, included virtual office hours and online discussion forums where students could ask and answer questions—and vote on which were important enough to filter up to the professor." Post-curiosity-launch-craze MOOCs are failing, especially if considered through the lens of the marketplace. In "Beyond the Buzz, Where Are MOOCs Really Going," Michael Horn and Clayton Christensen say, "In the current university system, for example, most faculty are rewarded for the quality of their research—not for the quality of their teaching. But MOOCs don't have quality teaching either. Students are one in a million." Online Design University may boost technical and entrepreneurial skills, but personal investment and interaction activate it in context, transforming Learning Skills into Making Work.

Onward: The truth is, medieval lecture halls are not empirically effective. College can evolve; a new system, designed from the ground up, can radically position Thinkers and Makers through an iron-sharpening-iron din of interdisciplinary collaboration. Education's objective is to educate, to abort the workforce cog molds. As the Education Business model exhausts itself, there will be room again for true discussion and interaction—a physical Sandbox in a physical playground with very present peers and live, engaging faculty, all reflective of contemporary design culture.

Interactive Isolation

Issues in the consumerist education model tie to problems descended from technological dependence. Students choose classes online exactly the way they order pizza from Domino's app. Specific advice based on professional and educational experiences can help flesh out the nuances of curriculum decisions, something not needed when deciding on cheese layers, so inserting faculty interaction and involvement is helpful. Harmful expectations of convenience stem from Tech and Money, and they manifest in similar means. In the case of Tech, the influence is somewhat indirect, as habits are set in a broader lifestyle sense before trickling into the student experience.

Much of this influence is detailed in Neil Postman's *Technopoly,* a dissection of America's particular embrace of any and all tech innovations. Postman's world argues that students are consumed—addicted—within a highly individualistic world of noise, distraction, self-sufficiency, and self-interest. Schools enable this behavior by sellingpointsofemphasis of tech vs. faculty, wires vs. people. This can manifest in headphoned design students surrounded by non-interacting peers. Under the guise of "interactivity," technology lets students draw deeper within themselves. This seems to be a non-controversial argument, yet it rarely inspires direct action because "technopoly, efficiency, and interest need no justification," writes Postman.

When not killing interaction, these tech-driven habits promote distraction. In multiple classroom crits, ras+e have told distracted phone users that "Facebook will wait until we finish," only to be answered, "You don't know my friends." This issue is profiled in the *New York Times* article, "Technology is Changing How Students Learn."Says Matt Richtel, "In interviews, teachers described what might be called a 'Wikipedia problem,' in which students have grown so accustomed to getting quick answers with a few keystrokes

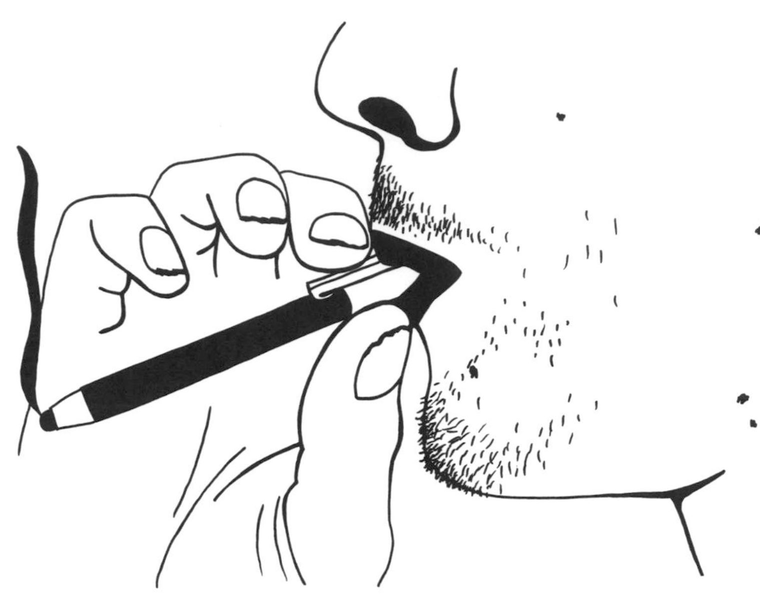

that they are more likely to give up when an easy answer eludes them. The Pew research found that 76 percent of teachers believed students had been conditioned by the Internet to find quick answers."

Capturing students' prolonged attention is hard, and it is not necessarily indicative of good teaching but rather a feature of entertainment. Hence, an easy overreliance on teaching through video games and TED. Pedagogy should acknowledge technology but should not cater to it.

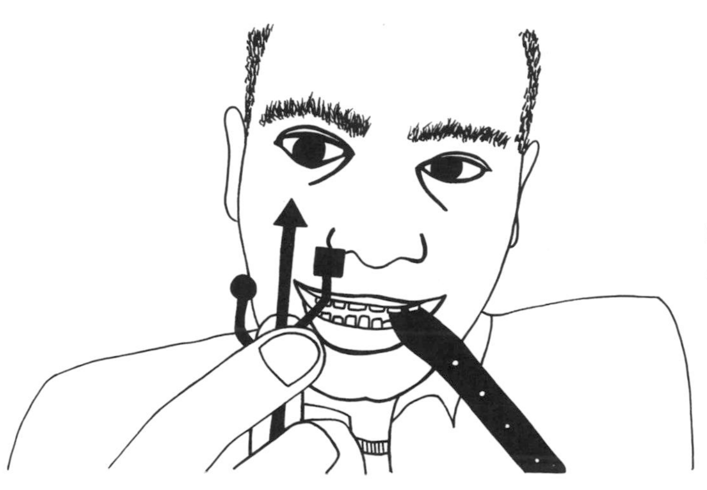

Copious use of technology doesn't have to become an inevitable distraction. According to *Proust and the Squid: The Story and Science of the Reading Brain* by Maryanne Wolf, we are not only what we read but how we read. Wolf explains how reading online is more like power browsing, and we tend to become efficient gleaners and decoders of information. Unlike deep reading without distraction, we remain disengaged and do not form rich mental connections. In design school, this translates into student difficulty in researching concepts and applying the information in an integrally linked way.

Students enter school, college especially, with unhelpful habits and traits already entrenched as a result of broader forces. Ludwig von Bertalanffy describes the biological and social impact of technology on mankind in his book *Robot, Men, and Minds:* "Precisely in affluent society, with gratification of biological needs, reduction of tensions, education and conditioning with scientific techniques, there was an unprecedented increase in mental illness, juvenile delinquency, crime not for want but for fun, the serious problem of leisure in an automated society, and the appearance of new

It's cheaper to change the student than the environment or the school

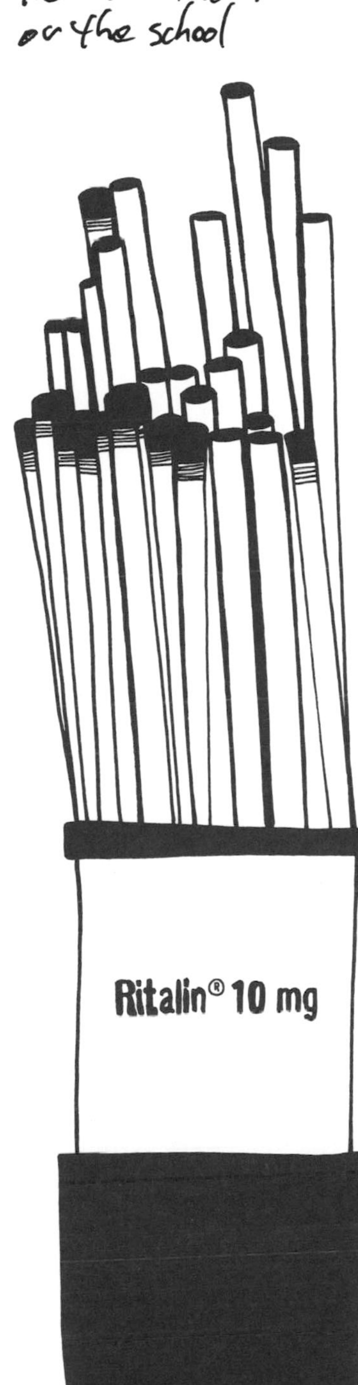

forms of mental disorder diagnosed as existential disease, malignant boredom, suicidal retirement neurosis and the like—in fact, all symptoms of a sick society."

If that all sounds like ADHD, there may be a link. Americans have turned to tech like a drug, a form of stress-reducing escapism. A need for constant entertainment in a hyper-connected world is an addiction. It has become standard for parents to ask doctors to prescribe Ritalin and Adderall for their children so that they can ignore the constant technological chatter and focus on work—this instead of the unspeakable alternative of monitoring usage. At least Hunter S. Thompson was able to channel his *using* toward production of work and away from isolation. Hopefully, designers can teach some other instructive methods.

Touch Tank

Creating interdisciplinary spaces, situations, tools, events, and coursework enables students to practice essential skills and create a broader range of work. Design created in physical proximity to peers is often collaborative by osmosis, and the pieces are measurably stronger than when students work alone, a potential missed in online interactions. This alone justifies the expense and investment in campuses and studios. Physical proximity creates new interactions—new tools—which beget invention. The physical structure of the building (or wireframe of the site), accessibility of materials, and layout of studios and buildings play major roles in shaping the atmosphere and voice of schools and programs that lead to productive interdisciplinary collaboration.

Facilities' structure and setup can impact workflow, communication, and vibe. Unsurprisingly, most computer labs feature sacred individual workspaces at the expense of interactive considerations. Monitors block sightlines to the projector, the professor, and other students. Lounge spaces are scrapped to increase class size. Locations of all gear and destinations impact how people move and what they come into contact with: printers, presses, darkroom, wood shop, library, cutting station, coffee, snack machine, music, bathrooms. Spacial convenience can increase the likelihood of students moving around and interacting; but deliberately forcing contact, even encouraging the likelihood that designers see what others are making, can build up mental databases of ideas and stimulants. Faculty should be accessed easily without blocking peer-to-peer contact. Studio computers can encourage users to drop in, but considerations for imported laptops are also needed. Furniture should not hurt bodies. The space itself should inspire pride in the work without conveying a hands-off museum quality. Spaces that students want to be in facilitate the "stickiness" of classrooms, programs, and entire schools.

Any studio located within a design/art school can be a Quad: students studying animation mingle with those pursuing information design, as ideas and skills jump from Maker to Maker. Tricks are conveyed and tech is absorbed. A senior working on kinetic typography for their thesis ends up collaborating with a musician.

Pedagogical reformers like Johann Heinrich Pestalozzi, Jean Piaget, and Maria Montessori stress the importance of play, curiosity, and physical environments in student learning. In "How a Radical New Teaching Method Could Unleash a Generation of Geniuses," Joshua Davis notes, "Einstein spent a year at a Pestalozzi-inspired school in the mid-1890s, and he later credited it with giving him the freedom to begin his first thought experiments on the theory of relativity. Google founders Larry Page and Sergey Brin similarly claim that their Montessori schooling imbued them with a spirit of independence and creativity." Even collaborative teaching is an effective, passive encouragement toward openness and experimentation, fostering a sense of live play. Referencing Black Mountain College as a creative hub, Bennis and Biederman note in *Organizing Genius* that "Great Groups often have a decidedly adolescent side."

Make Here

Makerspaces/Hackerspaces are proven effective incubators, and schools are adding them into the studio environment: accessibility of equipment and space (digital and physical) expand production, process, and interaction opportunities. Opening the spaces to teams and individuals outside the department—including business, community, and other programs—is essential. Suddenly, graphic designers can incorporate theory and production from industrial designers, engineers, hackers, philosophers, anthropologists, and other artists. Workshops, lectures, and various group meetings all find a home.

Such areas often deliberately eliminate boundaries between institutional spaces and the community. Baltimore Print Studios is a public-access letterpress and screenprinting printshop started by MICA husband-wife faculty team Kyle Van Horn and Kim Bently. Prompted by a demand for expanding the reach and facilities of MICA's printmaking building, Dolphin Press, while also developing the crossover of Kyle's printmaking and Kim's graphic design backgrounds, Baltimore Print Studios brings affordable self-serve studio access to Baltimoreans. Graduated alums and designers looking to expand their work also gain access to community and gear not afforded by their class schedule.

Hackerspaces can include physical and virtual environments, the latter being a network of users in virtual rooms who share resources, ideas, and projects. Corey Doctorow's *Little Brother* profiles four young students thrust into hacktivist roles when Homeland Security begins ruling San Francisco. The hackers distribute information by utilizing virtual meet-ups, digital pranks, and unregulated networks linked to gaming. In short, a digital educational institution.

Post-print, libraries repurposed themselves as medium-agnostic but, more relevantly, shifted their community roles to take advantage of their physical space and status as knowledge centers (e.g. Chicago Public Library Maker Lab). Various groups, events, and speakers shifted their activities into libraries, not unlike what NYC makerspace Eyebeam does. Similarly, 19-year-old Shawn Fanning's Napster became one of the largest and most important libraries in world history. The digital community and resource was born out of Fanning's time at Northeastern University, when he became frustrated with the limitations of his faculty and coursework, eventually dropping out.

Hackerspace.org provides a global directory of community-operated workshops where people can work together, fueling the argument that colleges are not essential for fostering productive interaction.

LAZY RIVER
learn and hang at Boston U

When it comes to school+makerspace+online, students reasonably question the essential qualities of studios. Napster was successful because students wanted free music, but the library aspect was the important cultural artifact. Mac Rumors attracts people who need answers to questions and hackers who are happy to give them. Reddit and Twitter provide an even more personalized RSS but with the ability for subscribers to contribute and comment. YouTube and Vimeo function as video libraries of expansive diversity. But virtual classroom hackerspaces are incomplete as *spaces* by not providing for physical interactions and osmosis. They also tend toward being stiff, difficult to modify and control, and are often redundant to the more casual pre-existing student worlds of Facebook and texting.

In David Kushner's *RollingStone* article "The Flight of the Birdman: Flappy Bird Creator Dong Nguyen Speaks Out," Nguyen explains how he removed his massively successful game from the Internet due to concerns about addiction. But it is exactly that world and trait that education game developers are pursuing. For example, INKids's Futaba Classroom Games offers educators the ability to plug in curricula-specific content into a game infrastructure that connects peers as competitive opponents. Schools are exploring the possibilities of interactive technology for play-based learning and collaborative socialization. Determining how much these tools teach or harm will likely come through future research, but regardless, physical studio classrooms retain important traits.

Physical contact in school matters. The telephone allows us to hear voices and email instantaneously transmits ideas, but in-person communication is fortunately not just about the clear, efficient delivery of information. Meeting with peers and professors promotes familial solidarity. Professors learn more about their students and are able to invest specifically, while students learn from one another's process. Group interaction, and coffee, keep everyone pushing through tiredness and setbacks.

Students who complete a MOOC are shown to have joined study groups that meet in person. In such all-skates, signing up is easy but completing the material can be demanding. The fun chatter and hum of a pub can be a more engaging and provoking space than a bedroom desk, according to Laura Pappano's *New York Times* article "The Year of the MOOC."

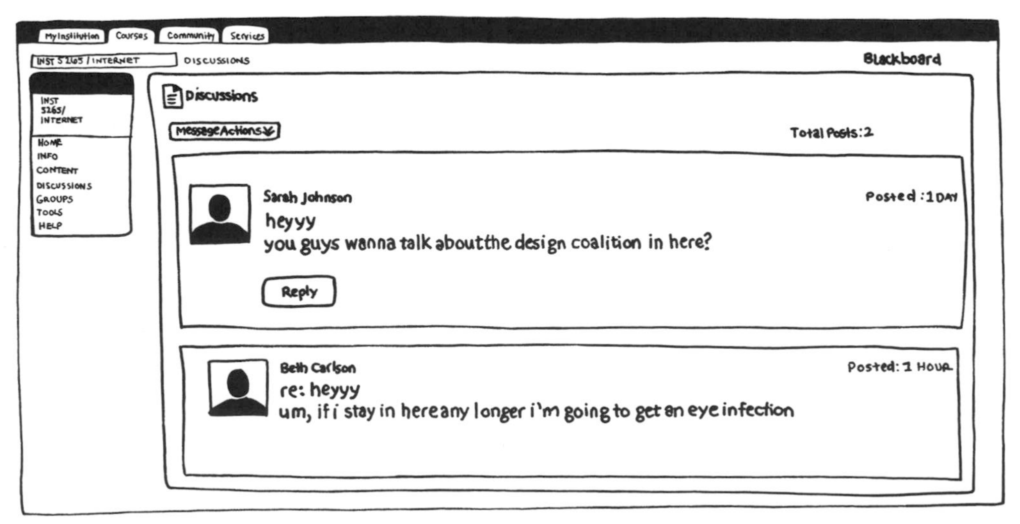

BLACKBOARD
the education software
for students to not converse

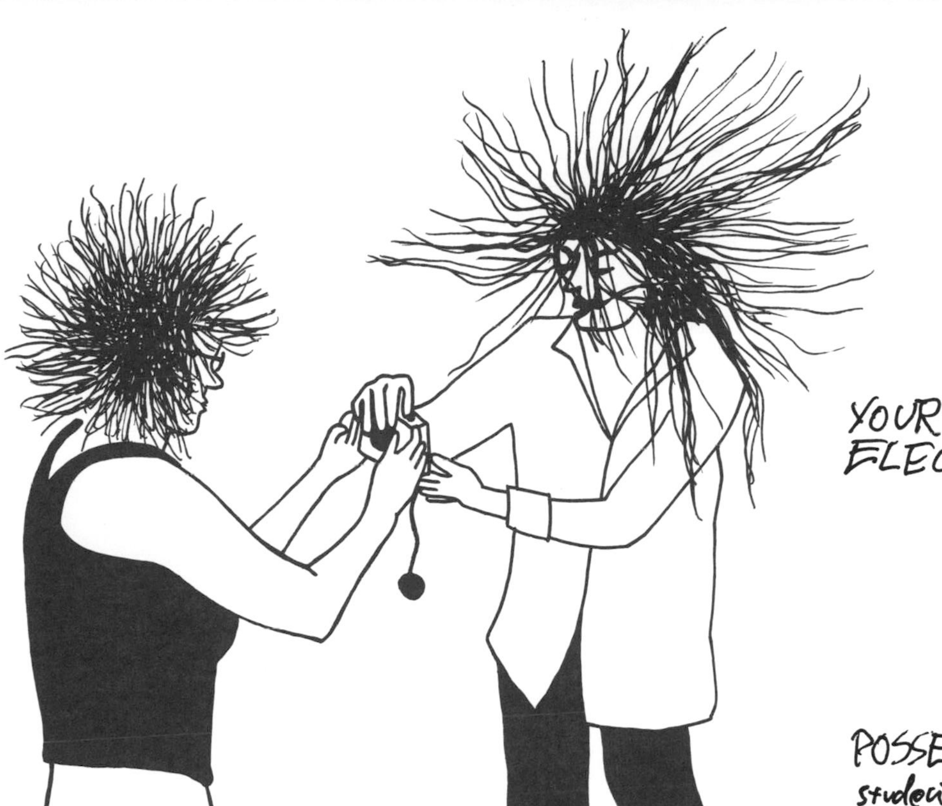

YOUR SHOTS ARE ELECTRIC

POSSE PROJECT
student support organization

In the *Wired* article "Virtual Reality and Learning: The Newest Landscape for Higher Education," Brian Shuster concurs, "At some point in their learning, every student needs personal help that interactive workbooks and textbooks alone cannot provide. Relying solely on asynchronous communication with a faceless professor stifles the kind of momentum that a classroom setting promotes. . . .The highest quality education must be social and interactive, and although online learning provides a degree of that via a website, the practicality of the real world instant feedback and social dynamics are missing."

Future design education models will ideally involve blends of digital and physical realms, anchoring online tools in a dedication to presentness. Parsing what *works* and what is *trend* will be crucial to maintaining deeply worked pieces and intensive collaboration.

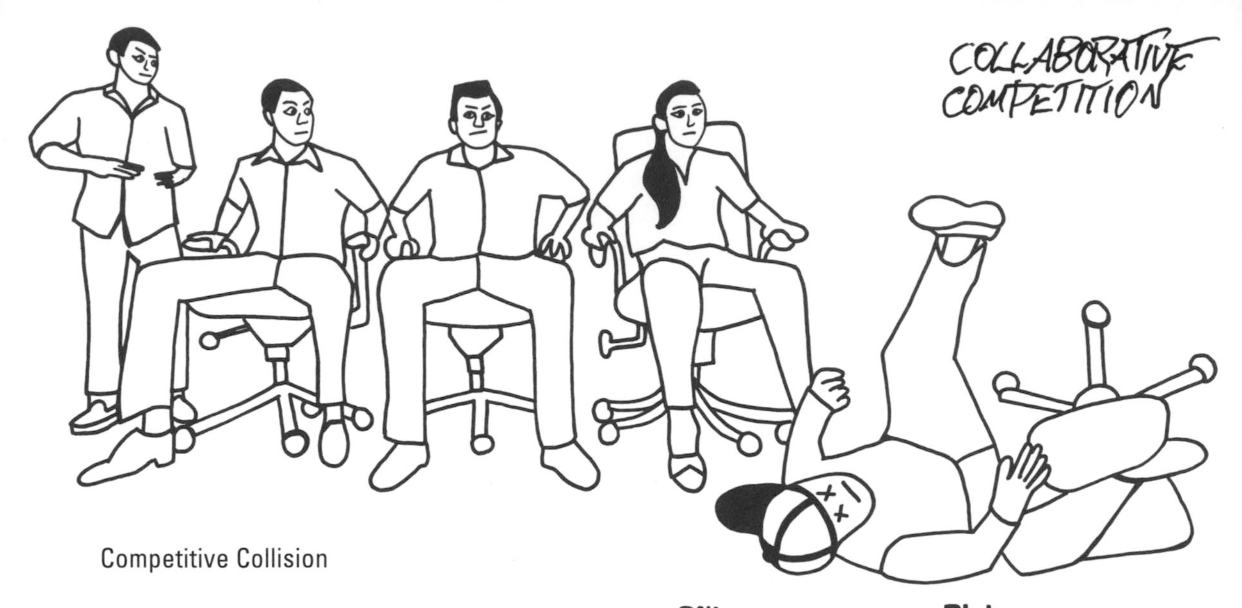

Competitive Collision

Administrative policies on diversity are often criticized for zeroing in on superficial definitions. But intellectual and cultural diversity based on life experiences, even deliberately incorporating non-design perspectives, are gaining traction. A group of students who do not know the boundaries of a discipline assume they have to work harder to catch up and are willing to experiment. The interdisciplinary sides of art and design practice demand cross-medium backgrounds and unexpected collisions.

Red Burns, Godmother of Silicon Alley, founded the Interactive Telecommunications Program at New York University's Tisch School of the Arts. The ITP ideology emphasizes diversity at the expense of technological expertise, selecting applicants based on student dynamics, which can vary from class to class. In the *Wired* article, "Let's Stop Focusing on Shiny Gadgets and Start Using Tech to Empower People," former ITP student Margaret Stewart writes, "Red wasn't that interested in technology *per se;* she saw it as something you needed to get to the real work: improving people's lives, making them feel more connected, bringing delight in big and small ways, and empowering them to affect change. …It wasn't a coincidence that Red created ITP inside NYU's Tisch School of the Arts rather than the computer science department; she wanted the program to be filled with dreamers, inventors, artists, and change-makers. … I sometimes describe ITP to those not familiar with it as 'Kindergarten for grownups,' but also love

another description I once heard: 'Engineering for poets.'… We may not know exactly what background or hard skills each [student] brings to the table, but we know we are likely dealing with an open, curious spirit; a great collaborator; and someone who is human-centered in the way he or she approaches problem solving."

Collaboration often derives energy from competition, a wholly relevant part of industry life. In the classroom, peers are simultaneously allies and competitors. Assignments are designed to provide high ceilings and open-ended challenges with room for students to experiment and produce portfolios that stand out. Collaborating supplies peer pressure and drives students to carry their weight and impress one another. Designers like Rodrigo Corral, Paul Sahre, and Project Projects create book covers within a narrowing print arena, inventively competing for eyeballs. In that world, formal mastery is assumed; idea is everything.

In *The New York Times* article "Do You Perform Better When You're Competing or When You're Collaborating?" Michael Gonchar details competition between children. American siblings are particularly competitive, even in mundane tasks like tooth-brushing. Competition is normal, innate, and healthy, providing an iron-on-iron problem-solving framework to expand. This occurs overtly in sports, both team and individual, a mirror to studio interactions.

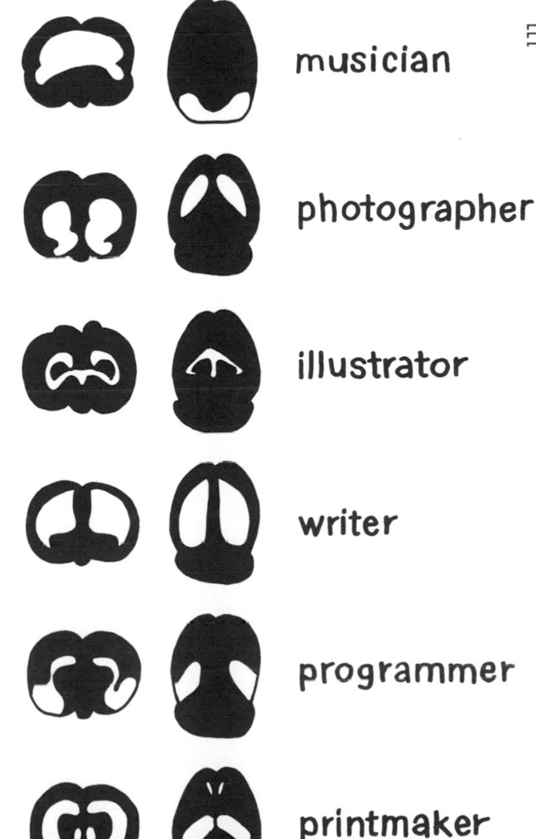

TECHNOPOLY

2. Free Range Design Curriculum
Student Lonerism

The American educational system prides itself on a Can-Do, individually customizable, *Choose Your Own Adventure* pedagogy. Human-Centered (What else would design be?), Student-Centered (What else would a class be?), guided only when necessary (Tutorials!). Yet, a general pall hangs over many schools that flattens out potential: a consumer-driven approach to education that reduces all perspectives to a job-training lens.

Not every college graduate prioritizes jobs for Thinkers: due to loan pressure, family concerns, or a greater interest in lifestyles detached from work, schools are pushed to prepare cogs. Americans fell victim to technological gluttony and a disposable culture, which trickled into program development. Overlooking the terms and conditions, schools went belly up, then compensated by recruiting rich Chinese families. Some schools are still criticized for deliberate, predatory programs that leave students with little portfolio POP but lots of tech know-how, large debts but little experience diversity. In the new studio world, that will not be enough.

CUBICLE
anti-collaboration furniture, not enough coursework is team-based

Graphic Design students typically work on projects alone; they chafe at collaborative work requiring creative flexibility. Neil Postman argued that technology encourages individuality; we see this in the resistance from students toward collaborative projects. Formal skills require personal practice, but collaborative skills receive significantly less attention. If job prep is so important, then why haven't design programs better prepared students for working collaboratively in a studio environment?

Despite the potential for unique combinations of talent and perspective afforded by college courses, the consumerist approach to education trickles into expectations that students have bought the right to avoid anyone's interests but their own. Design in a vacuum is not an accurate reflection of design reality. We agree that the Type I level is probably not the best time to implement large-scale collaborative work, but neither is introducing it for the first time at an internship. Employers do not *teach* collaboration; in an internship, students learn to work with colleagues through baptism by fire. In fact, design educators often hear complaints from design studios that there are a lot of talented applicants and hard-working designers graduating from top-notch design schools; however, getting these new designers to work in teams to execute projects is a struggle. Contemporary culture is driven by creative commons: sharing files, building upon ideas, and remixing images. Students, in our experience, often gravitate toward exploring ideas with others as a reflection of the online communities they take for granted. Courses and projects can be built to utilize these perspectives already at play.

WIKIPEDIA

FOLLOW INSTRUCTIONS ON TOP CARD

ALIBABA

PRICE $300,000,000

AMAZON

PRICE $300,000,000

CURIOSITY TAX
Florida fines students for taking classes outside their degree

WHY SO CURIOUS?

Many programs include New Media and Interactive Design in their Foundation program, and these skills and perspectives are essential. Students enter school with tacit experience in these mediums, so building context around cultural fundamentals from the beginning is hugely beneficial. Besides a taste of code, educators need to help students understand the dimensions of Interactivity and contemporary communication, its theory and history, exchanging and merging perspectives; this can push design and students' universes into a collaborative context from the beginning.

The Curious

The freedom to fail is widely considered a design process imperative, and school is a perfect environment to do so, or it should be. Going back to the Middle Ages, a master craftsman would adopt a young protégé. Imitating technical skills on the level of a specific technique is a lot like learning with training wheels. Today, many design educators still teach by this model. There is some merit in learning a unique way of working, which is why design programs run workshops with visiting artists. However, art direction at a curriculum level does not develop the ability to self-generate process and content, which demands freedom to pursue tangents and the freedom for those tangents to fail. Art and design students especially need to hear that it's okay to fail—that in our discipline, experimentation, the unexpected, and originality are way more valuable than competency or imitation. The most common answer to questions should be, "Try it and see." Interdisciplinary work demands curiosity. Curious George has a highly regarded place in collegiate studio practice as an impish role model.

As schools receive government pressure to graduate students in four years—for the economy, of course—faculty are less likely to encourage students to switch majors or add diverse classes that might increase student stay. In some cases, especially state schools, any additional classes students take outside of standard requirements are taxed with extra fees. This narrow approach to education actually deters curiosity, limiting the likelihood of inventive design solutions.

SOCRATIC INQUIRY
including of Design

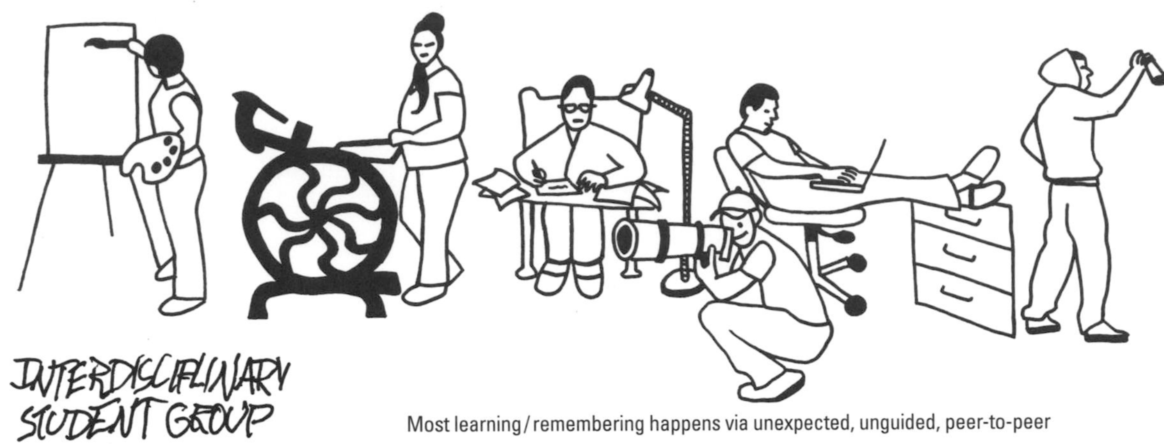

Most learning/remembering happens via unexpected, unguided, peer-to-peer interaction. Knowledge acquisition is a result of the collision between free people, free time, open space, and curiosity-fuel: "Hey, how'd you make that?"

Joshua Davis talks about free range intellectual curiosity in the *Wired* article "How a Radical New Teaching Method Could Unleash a Generation of Geniuses." His argument: young students are naturally curious, and education replaces that openness by training them to master very narrow approaches, limiting their potential for future discoveries. Davis details the work of a New Delhi scientist who let children explore a computer set up on the edge of a slum and discovered that they quickly taught themselves and each other without any adult guidance. In future work, Mitra is building schools with tech access but no formal teaching, letting the students organize themselves. For contrast, Davis discusses a 2009 study of children and toys, where conclusive test results demonstrated that children actually invent more uses for toys when nobody teaches them their use. *The Case for Make Believe,* by children's therapist and puppetry enthusiast Susan Linn, argues similarly that non-branded toys produce more vibrant, personal, and inventive storytelling when used by kids.

SWISS ARMY DESIGN
the interdisciplinary student

Design instructors need to separate ego from teaching practice, giving up tightly controlled art directing or idea-generation. It can be challenging to not "design" a course for a narrow set of skill-building outcomes, but not doing so allows students to learn more aspects of more topics, the cornerstone of teaching interdisciplinary collaboration.

Interdepartmentisciplinary

One notable attribute of design is that over the course of working with multiple clients on a variety of projects, practitioners learn something about many arenas, either through the content and subject or by constantly engaging new tools. Naturally, faculty are inclined to bring this perspective into their courses, breaking historical discipline silos as a reflection of their own projects.

The Graduate Communications Design MS/MFA program at Pratt Institute conducts a class called *Transformaction Design.* Changing every year, designated design faculty pair with faculty from another field, such as architecture or philosophy, to develop a unique hybrid class with crossover curriculum. Graphic Design MFA students at MICA explore writing as a design process with authors Elizabeth Evitts-Dickenson, David Barringer, and Ellen Lupton. At Parsons New School for Design, the academic division is labeled The School of Art, Media, and Technology, and their objective is for students to "explore areas of study and to learn how programs actively relate and converse with one another within Parsons' unique multiple-school structure—and with the wider New School network." The school prepares students by emphasizing the overlap between "art and design disciplines, expanding these fields beyond their traditional boundaries through interdisciplinary collaboration and exchange." RIT's School of Photographic and Imaging Arts also encourages innovation in artistic uses of technology.

FLIPPED CLASSROOM
or as design has always called it: "studio"

Increasingly, liberal arts and fine arts schools are encouraging faculty to explore interdepartmental projects and courses. The fine arts have always crossed art and writing with new media and graphic design tools, going as far back as the illuminated book of hours, to more recent iconics like Barbara Kruger. However, design studios and schools are fading the barriers between art, design, and technology as a reflection of cultural trends and the need for physical campuses to innovate, compete, and justify their price tags. Mediums rarely define careers, except when the mediums cross-pollinate in a distinctive way. Thesis students in American liberal arts schools are positioned to take advantage, putting together tools and techniques. Design students should encounter curricular crossover sooner than later. In his article "Interdisciplinary Courses and Team Teaching," James R. Davis writes, "In interdisciplinary courses, the faculty team members take on the chore of integrating their various perspectives and resolving their differences. In the ideal team-taught course, the faculty have successfully met the challenges of 'connecting learning' and the students have a chance to see the relationships that they don't get to see in other courses. This is one of the great pay-offs for inventing a new subject."

Flipped Classrooms

Collegiate faculty largely sidestep the waves of education trends, but the *flipped classroom* parallels the pre-existing studio ideal. Class time is used for exercises, discussion, and collaboration; while home time is dedicated to readings and lectures, made possible through online video distribution and communication tools. The most valuable educational resource—people—is maximized through student questions, dialogue, and complex problem-solving while allowing students to study at their own pace.

Accredited BFA studio programs run around five hours per week, most of that time being dedicated to practicing techniques and tools, discussion, and some group work. In-class open work time is required for collaboration with faculty feedback: skill and research–sharing, problem-solving, and informal critiquing.

Student-Run Studio

Student-run design studios offer "real world" projects, clients, a team context, and teach autonomy in making decisions while keeping faculty nearby to mediate professional problems, place student work into their community, and mitigate flaws in the internship model. If conducted as part of a course, the importance of grading or mandating equality of effort becomes subservient to the values and scale of collaborative projects.

Assigning a single, semester-long, intensive project forces students to manage a quantity of chefs while still protected by the safety-net of school. Students experiment with managerial processes and division of labor out of necessity. Leaders emerge. Often, students who rarely vocalize in traditional class meetings take on overt leadership roles, art directing and managing interactions with an Observer's experience. As professors dissolve their own authority, students maximize all available resources; peer pressure steps in. A rough semester breakdown:

Week 1: *Collaborative Curiosity* = Excitement + Planning
Week 2: *Clash of Visions* = Reality Check + Synthesis
Week 3: *Trial and Error* = Re-evaluation
Week 4–10: *Coerced Collaboration* = Groove Found + Production
Week 11–13: *New Collaboratives* = Mild Panic + Finalizing
Week 14–15: *Moment of Truth* = Dedicated students morph into feral leprechauns and roam the art building desperately in search of hot coffee and a pot of PVA glue.

A stand-alone course on collaboration offers students a range of relevant processes: new critique structures, swapping work, activist installations, forced connections, charrettes, sprints, and pecha kuchas. Some programs have loose structural definitions, allowing these perspectives to knit into multiple syllabi without adding courses.

3. Transactive Teaching

"There was possibly a time when show biz was a bigger business than education. Today, education is not only by far the biggest business in the world, it is also becoming show biz."—*War and Peace in the Global Village,* by Marshall McLuhan

Rockstar Teacher

Rockstars leading art and design schools have raised few eyebrows in the industry and education communities—in fact, big programs are expected to be led by big names. Learning from the masters is a long-standing tradition in the arts, and The Indebted certainly pray that supplementing studio time with elbow rubbing increases results, although those "results" vary widely for obvious reasons: Steven Heller leads mass at SVA in New York; Elliot Earls replaced the McCoys at Cranbrook; Ed Fella finally retired from CalArts; Kate Bingaman-Burt jump-started Portland State; Kimberly Elam stalks the rooms of Ringling; Yale is the podium for Mr. Rock. All invested parties—administration, faculty, students, parents—recognize at some level that many factors impact the many definitions of success, but that slot-machine lever has never looked so shiny as today, with graduate programs a booming business.

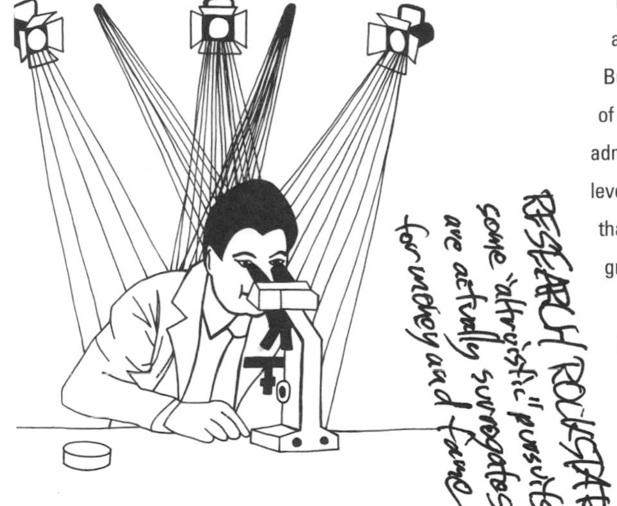

(Any criticisms of the system are in the context of the authors attending the Graphic Design MFA program at MICA under Ellen Lupton and Jennifer Cole Phillips, a context ras+e value immensely at every level. And speaking broadly, it is virtually impossible for any well-intentioned program to not benefit students.)

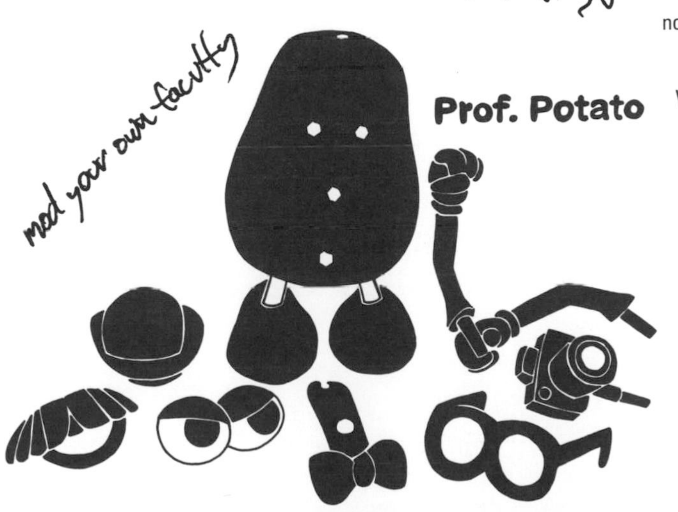

With students driving their own work in upper-level settings, faculty and peer input is always helpfully secondary to any self-determined results. The critique is with imitation; seeing aesthetic success and imitating aesthetic success amount to learning the wrong lesson. In attempts to please professors, many students often unknowingly emulate styles, adopt similar-minded design interests, and in some instances intentionally repeat the ideals and work of faculty. These issues worsen when the professor is a rockstar, well-connected, or otherwise seen as a gatekeeper.

Developing originality and problem-solving is more important for a student than successfully resolving any one piece. Some instructors have responded by hiding personal work or intentionally giving disparate feedback in order to not "rub off" on students. Having more teachers and more diverse backgrounds in a department is a common solution, accounting for programs avoiding alums in the hiring process.

Excavating Designosaurs

Recruiting rockstar faculty has an obvious advantage for a school, but the drag of the Outdated and Tenured Curmudgeon, grandfathered into (typically) small design programs out of formal respect for their old-pro work, makes responding to cultural and technical shifts difficult. They may be angry at the world because they feel that the design industry owes them something for their time and experience. With their value in experience, designosaurs' contributions mean much less in a world of rapid technology and culture shifts, where self-teaching and collaboration abound, a world of more Wicked Problems than vacant billboards. A school with a well-respected designosaur professor is kind of like owning a collectible car: its driveability is not really the point. Or, like how the Queen of England serves as a trophy on tour to rally support. As schools are squeezed for cash, a few faculty end up shouldering disproportionate loads regarding tech expertise. In a larger studio environment, 'saurs hire fresh blood to handle what they don't want to learn, but schools are rarely hiring faculty, and those they do hire are at low rank/commitment.

All an exaggeration, perhaps, but it is surprising that the consumer education model hasn't pushed the 'saurs out. The whole truth is that their experience is valuable—which quickly manifests in crits—even if they struggle in teaching the collaborative and interdisciplinary side.

Universities Love Collaboration, Have No Idea What That Means

Schools are leery of faculty collaboration. Even if they realize the potential and anticipated future reality of a discipline, co-teaching and co-making within an institution rarely have appropriate structures and viewpoints in place to assign credit, assess results, and supply resources. Mostly, schools are hesitant to pay two faculty for one class. Considering the individual nature of American education, this issue is unsurprising.

Common university grievances include: committees evaluating similar research packages, co-teachers grading assignments and providing feedback from multiple perspectives, and overlapping allocation of time and energy. For these reasons and others, collaborating faculty may be perceived as slackers. Some of these issues stem from the insistence of the Education Business model, and some come from narrow viewpoints. The arguments for co-teaching are precise: if design and culture reflect trends toward collaboration, then education should reflect this also, and the second argument simply values diversity of input. Certain topics are best experienced through alignments of particular backgrounds and skills.

Penn State Anthropology Professor Dr. John L. Jackson Jr. attacks the misconceptions of faculty collaboration in the aptly named *Chronicle* article "Co-Teaching is More Work, Not Less." Collaborating requires immense investment of faculty. Synthesis is usually hard and typically messy. Creating courses, assignments, and lectures and determining what is relevant and to what degree—effective co-teaching requires all this to be done as a reflection of two points of view, and maintaining a tight hybridization throughout a course requires constant inter-faculty communication. Every decision is made individually and resolved at Round Two. Teachers do it because they want to, because they're curious, and because the deeply layered results favor students.

Fill the Bubbles vs. Connect the Dots

Students have access to all the information they could ever choke down. That's before they attend college.

At a visiting artist lecture by Information Designer Jer Thorpe, a student critiqued the design curriculum: "We are taught how to make information pretty, but we're not taught how to understand it." Without context, representing information is just Style. A non-critical judgment of large amounts of *stuff* is U-Haul/PODS storage facilities, American obesity, and *Ripley's Believe It or Not.* With quality control, eight different social media subscriptions wouldn't all chirp at once. The greatest service a professor can offer is context and perspective on the info/media onslaught.

To collaborate, students must first connect to one another.

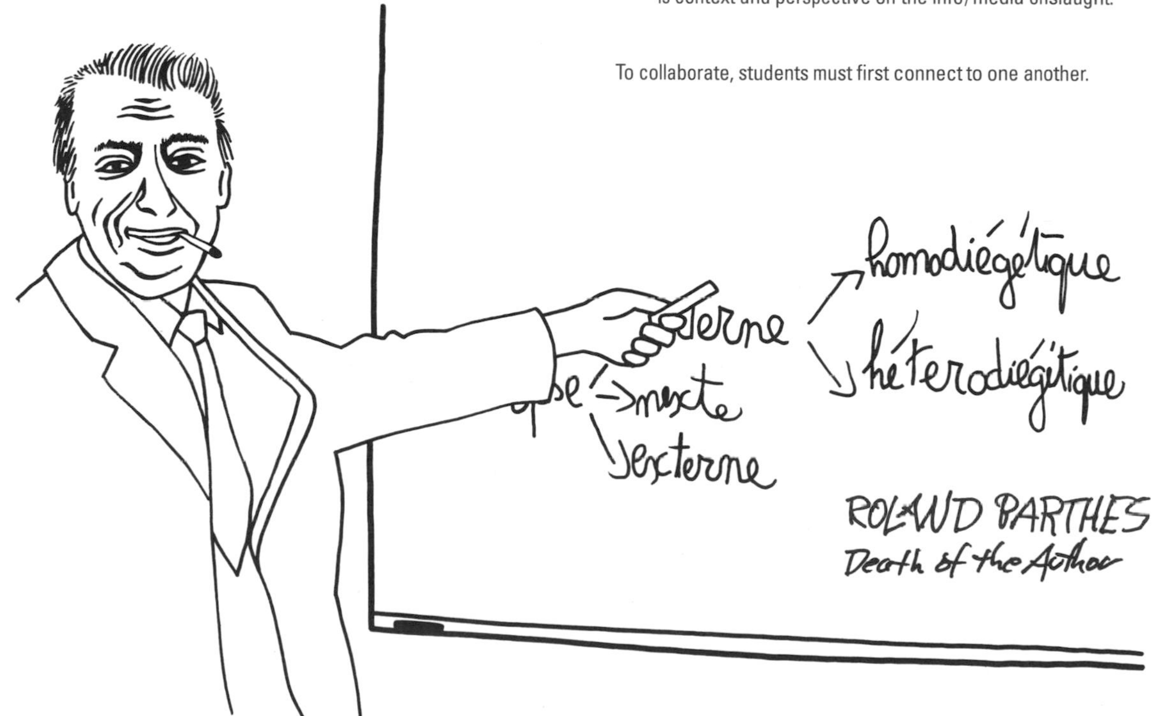

Scene: A professor stands in front of a vinyl screen with a projected image of William Morris. A student faces a lab computer and logs onto Facebook. For those who weren't listening, the Arts and Crafts movement aimed to bring Art into Life. Click. The Bauhaus appears. Laszlo Moholy-Nagy, Constructivist and Instructor, pushed humanism within the framework of an industrial society. A phone vibrates and another pair of eyes diverts. The layered walls of screens in the classroom provide info, or not, based on the whims of shrinking attention spans. The fact that students would rather be on social media and drift in and out of a lecture is not surprising; it used to be comic books hidden in textbooks, but the result is roughly alike. Presentness is important, curation is important, the ability to focus deeply on pushing solutions past surface is vitally important. These are design skills, industry skills, and student skills. Faculty bear some responsibility on proving the imperative of parsing relative importance. Students need to know what Will Morris does for their reality.

Education as entertainment can disguise learning, conflating the value of something with how *fun* it is. This matters in life, but it matters in design because the best solutions are often buried beneath the obvious. It's not enough to solve the problem, the result must be interesting, inventive—and it's a competition.

While the goal is not engagement through entertainment, collaboration requires students to be present. It's an uphill battle. Contemporary high school is anti-art, regimented bubble-filling, and students have to juggle the openness of college studios in addition to the trial and error of invention. Students need faculty help to connect the dots, to assemble something new from the pieces they're given.

"According to Friend & Cook (2003), parity means that each person's contribution to an interaction is equally valued, and each person has equal decision-making power. Gately & Gately (2001) note that co-teachers may have to progress through a compromising stage before parity develops."-*Shared Insights from University Co-Teaching* by Greg Conderman and Bonnie McCarty

Transactive/Co-Teaching

Transactive teaching means collective instructing, and it naturally engages students in material through constant context and juxtaposition, replacing monologue with dialogue. Sometimes there's even collision. Perspectiveclash provides accuracy, like how paradox is the best truth. Students watch things being hashed out; they see multiple angles lead to multiple solutions. Co-teaching in two-person teams is most common, although workshop courses, in which a single voice for a condensed period of time rotates out with other faculty later on, can be effective. Quality investment in communication is an imperative, as live response needs to reflect an agreed-upon duality. If students sense conflict, their trust in the experiment is over.

While co-teaching typically involves more work than teaching solo, there are advantages in the quantity and depth of the material that can be covered, since two bodies of experience are available. Physical classroom feedback is also quicker with two sets of feet roaming the aisles. In a culture plagued with automated responses, increasing personal/custom contact time is something to be embraced. When the faculty are confident in their ability to represent a shared vision, when parity is achieved, answering emails and grading can all be split between two sets of hands. The roles and division of labor within a design studio course can vary greatly, with faculty able to break classes into smaller chunks, play to lecturing/critiquing/demo-ing strengths, divide content by area of expertise, trade duties, or simply roam the room.

When courses are structured to be overtly interdisciplinary, usually with a Foundation or Special Topic curriculum, the faculty bring their shared disciplines together. In a Foundation framework, the diversity helps to build a broad base. Special Topic courses allow two different programs to collide, such as typography students working on visualizations with poetry students. Almost all Wicked Problems or Social Design coursework rely on a range of student disciplines, and bring in either guests or co-teaching faculty, since the topics cross between many arenas.

Fears and Cheers

Concerns from students about team-taught design courses usually stem from a combination of assumptions and lack of dialogue with the faculty. Instructors need to make sure expectations are clear and consistent, first and foremost. Students need to adjust to the unconventionality, and teachers should be aware what they are doing is experimental. As is true with any collaborative work, a willingness to buy in and a determination to contribute with humility are absolutely essential.

The greatest concern is that students might receive conflicting critiques of work or, more likely, differing advice on moving forward. However, multiple viewpoints effectively counter one-sided thinking and stylistic imitation. Students learn to assimilate all wide-ranging advice into work that reflects synthesis with their own perspective. Having multiple inputs reinforces the openness of possible solutions, allowing room for the student to decide and innovate.

SECTION

02 Case Studies

In Q+A form, a broad selection of practitioners across a range of disciplines describe their collaborative process. Consensus and discrepancies emerge as designers, musicians, educators, architects, and artists tell us why they choose their mediums and how they work with others.

Project Projects

Rob Giampietro, Principal
New York, NY

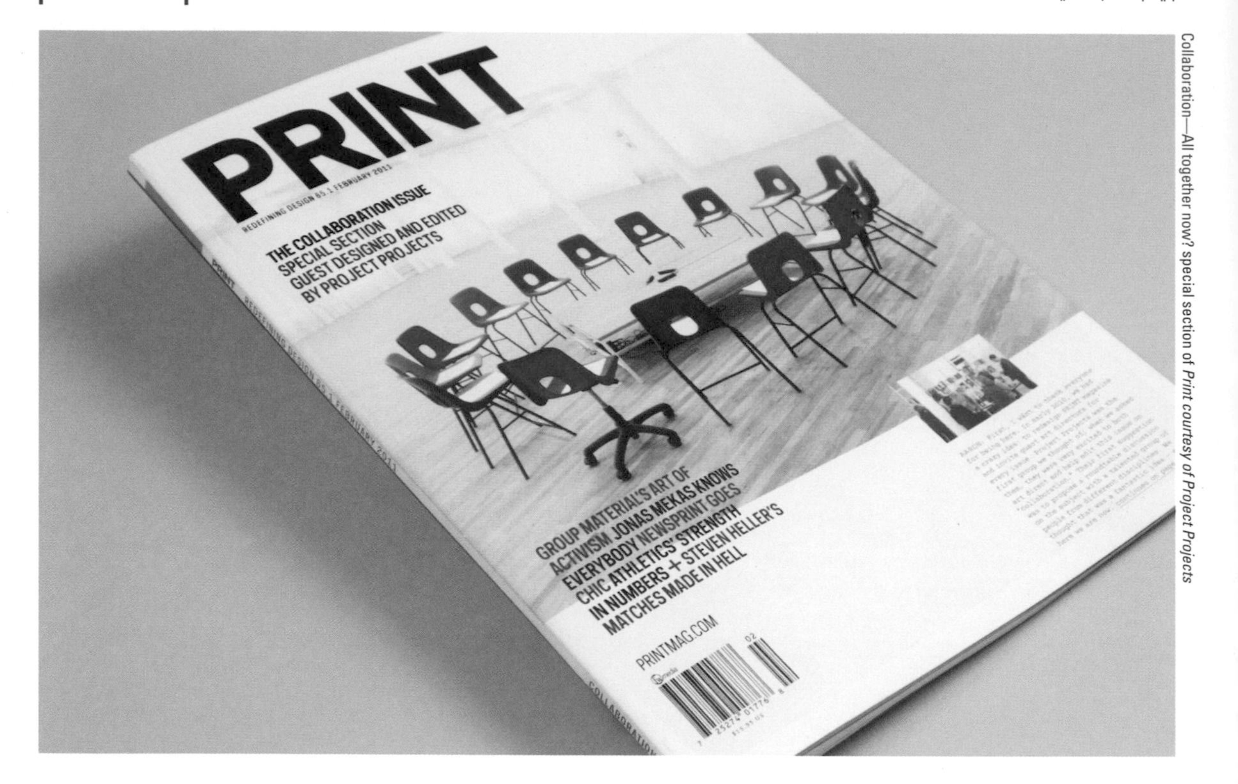

Collaboration—All together now? special section of *Print* courtesy of Project Projects

Why a special issue on collaboration for *Print*?

Print came to us. They just received a new editor, and were rethinking how they were doing things. I'm sad that they are gone now, but at the time they had this idea to commission different designers to do the middle section of the magazine. They thought it would be great to kick it off with collaboration, because it would be a collaboration between the magazine and us, as the designers. Also, *Print* viewed our studio as being particularly interesting in terms of the ways that we collaborated. So we wanted to get a bunch of different people from different areas together and talk about collaboration. We also thought about approaching different writers that we knew, and we had story meetings. Project Projects turned into a little editorial group during that time. It ended up being kind of fun to engineer these opportunities, and also to commission photographers, and work with all of the type designers that we knew to give us an unreleased typeface for the article headlines. It wound up, in a way, being a portrait of our network of people, some of whom are designers, but many of whom are not. It was really fantastic to also work as editors, which is one of the founding interests of the studio: how design and editing feed back on each other. We wound up having this panel discussion that was private, but that we recorded and transcribed, and we let it run throughout the issue. One thing that's cool to see is how that little transcript pairs with things that are being spoken about within the article that is hosting it. These simultaneous texts rhyme and clash at certain points.

Is writing a collaborative medium?

In some ways it is very un-collaborative in that you are often by yourself, hammering away at something. But at the same time, you are in dialogue with other writers that you have read. Almost every piece that I wrote for *Dot Dot Dot* came out of numerous conversations with David Reinfurt and Stuart Bailey. We would just be talking about a set of ideas, and then at a certain point, one of them would frame the idea, "Can you write a piece about that idea for the next *Dot Dot Dot?"* It always came as a surprise. So you go off and try to figure out what you think. There is also a sort of dialogic collaboration that happens with the editing process. I find that while writing is very similar to speaking, designing collaboratively is really about listening. The longer I listen before I speak, the better the solution is often going to be.

UC Quarterly publication of the UC blog network's greatest hits *courtesy of UC*

What is the Armin+Bryony collaborative process?

Armin: It's both minimal and maximal. We divide tasks. I'm usually in charge of the designing aspect and the building of our websites. I work double Bryony's hours because she takes care of our kids in the afternoon. So we rarely sit down to "brainstorm" or work hand-in-hand on projects. When it comes to design, I'll do a few options, then show them to her, and in three minutes we have a decision to move forward with something or not.

Bryony: We know when the other is slacking, or when things could be better, and we don't bullshit or tip-toe around it. Having said that, we allow each other a lot of freedom, and we really try to do things that are a bit risky and different.

What about co-managing your studio?

Armin: Bryony is in charge of the business and logistics of our events and competitions. I'm in charge of designing and producing. We've divided our work/home lifestyle so that our kids are not shipped to daycare 12 hours of the day, but instead come home at 1:00 pm, and Bryony, who is a great mom, takes care of them. Also, since much of our work resides online, and I'm the web person in the relationship, I have more work duties.

Bryony: Our dividing tasks comes naturally by taking advantage of individual strengths and realistic expectations. I would love to have more time to design, but it is hard to really dig deep creatively when your longest uninterrupted chunk of time is three hours. I stopped giving lectures when the babies ruled our house—now that they are older I can leave for a few days and the world does not fall apart. We often hear people say they don't know how we can produce as much as we can, and the simple answer to that is that not a minute is wasted.

How do you compliment each other?

Armin: We have very similar ideas of what's good or funny or interesting. Bryony is very crafty and good with her hands, whereas I can't glue two sticks together to save my life. One example was the identity for the 2013 Brand New Conference, where I had designed a pixel-based logo and a background pattern that was very clean and computer-y. For the program, Bryony suggested we take the logo and sew it onto each cover, by hand. I never would have thought of that solution, mainly because I couldn't imagine actually executing it.

Bryony: It helps that we have the same goals. We are like a pinball machine always bouncing forward.

The Art Guys

Michael Galbreth & Jack Massing
Houston, TX

Yard Crew guerrilla event, Contemporary Arts Museum, Houston, TX *courtesy of The Art Guys*

What is The Art Guys process?

Mike: It happens different ways and changed over the years.

Jack: It derives from discussions, a long history, and a long collection of ideas that we've batted around. We happen to be fairly similar in our humor, our interests, and our desire to make things that are not necessarily Art, although we fall into the category of Art. We believe that art is something that is incredibly important in the bigger scheme of things, and that everything resides inside the idea of art. The collaboration is something that occurs in every shade that you could think of—between 100% of Mike's doing something to 100% of me doing something. Sometimes we do whatever we want to do, and other times we really bat stuff around and go 50-50.

Mike: There are two of us, so if there are profoundly bad mistakes, it means that two of us have fucked up. Generally speaking, any problems that The Art Guys may make are often mediated before they come to life. Early on within The Art Guys structure, we called our philosophy of working *The Theory of Wrongheadedness,* wherein we would choose to do the wrong thing at the wrong time for the wrong reasons and use the wrong material. Sometimes things can be so wrong as to seem right. There's something about that. All of this is just thinking about the world. Why is it that we, as artists, have chosen to do this rather than something else?

Can you talk about your individual skill sets?

Jack: We bring our own stuff to the table. Mike is unbelievably intelligent, very receptive, and efficient. He gets things done very directly; I fiddle around too much, but fiddling around can become very fruitful. When we are faced with the task, we both hit it head-on and knock it out on time and under budget. We often comment to each other that we could open up an ad firm or a design firm. We can do so many different things as a team and be successful, but managing "artwork" has been really interesting to us. I think it is a little more difficult and a little more risky than producing a product to be bought and sold.

Is there a piece that reflects two perspectives?

Jack: They all have both of us in there. I suppose that our performance work could probably reveal more and be understood best that way. When we are on stage, you can actually see us doing what we do. Although we decide beforehand what we are going to do, we don't necessarily rehearse. Because we don't rehearse, we are not really

SUITS: The Clothes Make the Man year-long behavioral event *photo: Mark Seliger, courtesy of The Art Guys*

performing, we are behaving. Behavior on stage may not be very desirable for an audience. That doesn't necessarily matter to us; we are coming at it from a sculptural point of view where you have a task and you do it. In a way, it is a Fluxus idea, where you come up with an idea conceptually and then perform it. The performer is not necessarily the focus of the audience's attention; it is the idea. Unfortunately, people who are sitting there in a comfortable seat are expecting to be entertained, and that is something that we like to mess with.

Mike: Most of the public event things are much more effective than other works of ours because there are two of us. For one person to do something that is seemingly stupid is one person doing something stupid. But when you have two people doing something stupid, it is more than that, because it implies a communal activity. I think it welcomes people into the dialogue. Projects that we do in public, the so called "out in the streets" things, work better as collaborative things.

What project was particularly successful?

Jack: We did a project where we leased advertising space on our suits for a year. It was a couple of suits that we wore around, and it became iconographic. That piece is so nuanced and valuable to us in many ways because it had many different elements in it, and we tried to derive an income from it. We tried to make it interesting socially, physically, and in a performative way over a year-long span. It took us a year to produce it before we started wearing it. We spent about a year cleaning it up, and making a documentary and a book, so all together it was about a three-year project that resulted in so many different residual projects, or ways to understand and view the work. We had to be performative. We had to wear it around every day and clean it and talk about it endlessly while we were wearing it. We were dealing with the media, and it all worked to our original vision.

Do you have an example of how the collaborative process enabled you to resolve something?

Mike: I remember one difficult situation early on, and it had nothing to do with any conflict between us. We had proposed to work with a temporary public art situation, and it was rejected for very poor reasons. I was ready to walk away because I didn't need that in my life, and I didn't care if they didn't like it. But Jack is much more of a conciliatory person and he said, "Why don't we just come up with another idea?" and it worked out well.

You learn a lot from working together. We usually have a pretty good sense of whether something is bad or if it is good in the moment. As far as the work is concerned, there is usually not that much disagreement at all. For example, we are often asked to be judges for student exhibitions, or to choose works for something. We can go around the room, and even though there may be 100 works to look at, we are done in five minutes. As judges for a show, it is funny that we both invariably agree exactly on what works are the best, without consulting during the judging process; we often don't talk to each other while we are judging. This happened recently. The process for judging this show was to make a list, ordering the best first and the worst last. And by God, our lists were identical. I don't think that is an accident. It comes from clear thinking and recognition that some things are better than other things.

Why the emphasis on humor in the work?

Mike: Neither of us are particularly sad people. The Art Guys have always been doing things in public, and when you do things in public, you want attention. You want a dialogue. Humor can often be a welcoming thing. Once you have their attention, then you can do what you want. You can engage them in the ideas.

Jack: Slapstick always seems to work because it is not language-based. I think that has a lot to do with art too. The juxtaposition of things is universal.

The Questa Project

Martin Majoor & Jos Buivenga
Arnhem, NL

Questa Grande + Sans + Regular *courtesy of Majoor+ Buivenga*

Majoor (L) + Buivenga (R) typical work session

corrections forgotten ownership of red + blue pen

THE QUESTA PROJECT

Questa

Cactus siesta
useally at 14:00 hours it is time for
The office
Small Caps Lock*
Industry standard input and output

RÉSUMÉ
(Modern) American Usage is allowed
One hydrophore
The *basic* idea from the FBI was...
aquamarine?

Questa Sans

Cactus siesta
useally at 14:00 hours it is time for
The office
Small Caps Lock*
Industry standard input and output

RÉSUMÉ
(Modern) American Usage is allowed
One hydrophore
The *basic* idea from the FBI was...
aquamarine?

Questa Grande

Cactus
& 14 hours away
The office
Small Caps Lock*

RÉSUMÉ
(Modern) American Usage
Hydrophore
Aqua!?

www.thequestaproject.com

The Questa Project Poster features Questa Regular, Sans, and Grande *all images courtesy Majoor + Buivenga*

How do your backgrounds mesh together?

Martin: We have a common background in our art education, attending the same school in Arnhem in the ’80s and meeting occasionally. After graduating, we went our own ways and didn’t meet again for about 25 years.

When it comes to type design, Jos is self-educated. He worked in advertising, but he admired the type designers he knew from Arnhem. I had the opportunity to study type design, and I worked as a book typographer and type designer after graduating. Since we started collaborating, our technical skills and aesthetic insights have balanced, but we still learn a lot from each other.

How do you work together?

When we started Questa back in 2009, we both lived in Arnhem within a stone’s throw of each other. It was very important to get together in the same physical space—especially in the first few months of brainstorming, sketching, and printing. I regularly traveled to Warsaw, so we started working remotely through Skype’s desktop sharing option. It was not the ideal situation for collaborating, but we made it work. Meanwhile, Jos has moved away from Arnhem, but we see each other almost weekly, in his well-equipped studio.

How do you hammer out details when working?

We differ in our approaches. Jos once put it like this: “Martin tends to be more experimental and I am more likely to be satisfied with traditional solutions.” Jos has a lot of marketing experience, whereas I know more about the history of type design. All of these aspects help us make the best possible decisions in our collaborative process. We sometimes split production, but everything concerning the design process, we solve together.

Is mixing sans + serif a collaborative metaphor?

Not long ago, serif and sans had their own background; they were almost considered ‘enemies,’ unable to work together on paper, or at least not without difficulty.

Questa, however, demonstrates that when deriving the sans version from the serif, they can mix seamlessly. At the same time, both serif and sans keep their own identity without being competitors. Extending the metaphor, Jos and I have a common background and collaborate closely, but we always keep our own identities.

TypeTogether

Veronika Burian & José Scaglione, Co-Founders
Spain & Argentina

KARMINA. POSTER SPECIMEN

Japanese tea

Bereits um 565 wurde es erstmals schriftlich erwähnt.

beautiful texture

TESTING THE BOUNDARIES

534 live lobsters escaped the dinner plate and belly flopped to freedom

calligraphic reminiscence

quält jeden größeren Zwerg

“Kö”geht nicht

Im Jahr 1934 ging ein Bild um die Welt

neutrality

Karmina type sample poster *all images courtesy of TypeTogether*

O prázdninách měl Martin Pechlát napilno – celý srpen hrál každý večer na Hradě dvojroli Antifola v Komedii omylů a přes den zkoušel ve svém domovském Divadle Komedie náročnou roli v inscenaci Goebbels/Baarová. Navíc ještě stačil natočit film s režisérem Davidem Jařabem. Zdá se, že ho žádný úkol nevyvede z míry tak jako rozhovor. V civilu je totiž plachý introvert...

JANA SOPROVÁ

VĚŘÍM, ŽE BAAROVOU OPRA[...] MILOVAL

Karmina in DN divadelní noviny design: Designiq

Abril in paul-rand.com design: Daniel Lewandowski

Abril Fatface + Bree Serif *courtesy of TypeTogether*

Do you have studios in Spain and Argentina?

Veronika: Yes, we both have a studio in our respective houses. Our employees are in Germany, England, Greece, and Argentina; and we have more collaborators spread around the globe. We collaborate remotely through Skype and email, and we use online project management software to keep track of our projects. We try to talk every day, and we are both involved in every decision.

The company is registered in Prague, but we do not have a representative studio there.

How often do you meet?

I meet with José about two or three times a year, and we see the rest of the team once a year at AtypI.

Does proximity impact your process?

The actual process does not really change, but it can be more intense in person. However, we mainly talk about other parts of the business and new font ideas. We do not do heavy design or font production work, which is so time-consuming, because it would be a waste of our time together. It is helpful to be in the same place when developing ideas for a new design and to give feedback on drawings to our team. We make a conscious effort to plan our projects with a long-term view of our library.

Are type design studios trending towards smaller, interdisciplinary collaboratives?

Yes. The democratization of tools and commercialization of online-typography allows small operations to exist. Indie foundries are a relatively new development, and they bring a breath of fresh air to the industry. The increase in the publication of new quality fonts over the past few years has made it harder to be a one-person foundry, if you expect to make a living out of selling your typefaces. We think it is more viable to join forces and create a type cooperative, which we see happening a lot.

Do you work on one aspect simultaneously?

We can work on any aspect of a font, but the rule of thumb is that we both have to look at everything at least once. We trust each other and do not feel over-protective of our “type curves.” It is essential for both of us to keep the other person’s opinion and eye in mind. Also, it’s important that we share a similar view on our profession, and that we are both critical of our own work.

Katherine+Michael McCoy

High Ground Design
Buena Vista, CO

How has collaboration in education changed?

Katherine: The National Association of Schools of Art and Design (NASAD) rewrote their standards this past year to include a larger emphasis on collaboration. I am one of their trained "accreditors" for site visits. NASAD recognizes that design is becoming more complex, and collaboration is a necessary method of working. Projects are becoming very large and span many different disciplines, so teaching students collaborative techniques is essential.

Michael: Within industrial design and product manufacturing, it has always been a collaborative process, whether people want to admit it or not. But if they admitted that it is a collaboration, then they'd be able to use tools and methods to enhance it, rather than make it painful.

How do you teach collaboration?

Michael: In studio courses, selecting the right project is important. Teachers have to assign a project that is complex enough where they actually need each other's skills and insights. It's about understanding the moving parts of a project, and which points require different skills. Then students have to agree on a theme, philosophy, and approach. Along the way, everyone can do their own thing at the appropriate moment. Eventually it comes back together, and fits as a whole, as opposed to having fragmented, disassembled parts. From the beginning, students should understand that they need each other, and that they won't be doing identical things. Each student brings something different to the table. In the past 20 years, Kathy and I have focused on storytelling and narratives as a way to get everybody involved in collaboration.

Katherine: Those methods seem to ease and stimulate the collaborative process. Over the past decade, we have been conducting High Ground Design studio workshops and studio conferences that teach collaborative techniques. When putting teams of strangers together at our workshops, the methods of storytelling and literally acting-out design solutions help people break down social barriers. Especially in the early stages of a collaboration, storytelling is an idea-generator. It helps people in groups of diverse disciplines, skills, and personalities, because everybody knows how to tell stories. Storytelling is an integral part of human development and culture.

Michael: It's a problem especially with complex industrial design projects. Engineers, designers, and financial people are pulled into that first meeting, and everyone is nervous because they speak a different disciplinary language. We think it's important for groups, because it flattens out their hierarchy.

Workshop Storytelling + Timex Design Staff scenarios + acting-out are interrelated co-lab tools *courtesy of McCoys*

What objectives do people have when they come to the High Ground Design Workshops?

Michael: A lot of them work on teams in organizations or corporations. They have to produce some kind of product or service. A poor collaborative team has multiple voices tugging in different directions, so our guests want to learn how to work better as a whole, while not succumbing to the unfortunate design-by-committee phenomena. The worst kind of collaboration is one that's really bland because everyone beats each other up to the point where they all give up.

Katherine: Yeah, it's like that saying: *A camel is a horse designed by committee.*

Fortunately, the business world has discovered design thinking in recent years. One of its key elements is conceptual synergy, and that results from successful collaboration. Many of our workshop participants are not designers. We find people that are interested in learning the tools that help collaboration.

Yale CEID

Joseph Zinter, Assistant Director
New Haven, CT

students at work in CEID makerspace *courtesy of Yale CEID*

What made the Yale School of Engineering decide to invest in a non-degree-granting makerspace?

Working on real-world and open-ended problems in education has gained a lot of thrust over the last few years. In the engineering sphere, a lot of people are beginning to understand the value of bringing many different voices into one room. It's a technique that designers have been using for a long time to solve real-world, open-ended problems. Engineering spaces are pretty good on the studio/shop facility, but not as robust on what it means to innovate and make. I had an urge to build the engineering sphere of a design space. A number of spaces are popping up in many different schools. It's something that I think academics and universities are realizing the value of, because of all these different makerspace models that are demonstrating successful work.

What is the Yale CEID set up to do?

We're basically conducting an ongoing experiment. We do a lot of interdisciplinary work. The CEID occupies the former engineering library. Thus, we've mixed books with tools. There are spaces for collaborative making, a classroom, and multiple conference rooms. The space is reconfigurable and reservable by students, faculty, and staff. The large studio covers about 50% of the space, housing everything from Styrofoam balls and Popsicle sticks, to a sewing machine, computer workstations, power and hand tools, electronics, and 3D printing. We have three other workspaces—a metal shop, wood shop, and wet lab.

We also do a lot of programming. We support students teaching students. A number of student groups put on their own programming and courses in an extracurricular type of way. We have a lot of social and networking events because community is a big aspect of what we are trying to establish.

The response from the community has been mind-blowing. We have about 1,400 members. Our members represent 58 of the undergraduate disciplines. A diverse group of folks is exciting.

We are situated on one of the busiest stretches of campus, which is fortuitous. And we have a 2,000 ft² window, which looks right out to the street. When we set up the space, we were very conscious of putting the interesting and exciting elements of collaborative production right in people's faces. People walk down the street and think, "If that's what design and engineering is, then I want to be a part of that."

CEID is about demystifying technology. We've got projects going on with MBA students in the School of Management. We're working with people from the Museum of Natural History to generate 3D-printed models. We're working with folks from the School of Art. We're working on engineering research. We're doing a lot of work with the School of Medicine. In a tangible way, we are really becoming the school hub.

Why are 3D printers popular across disciplines?

The barrier to entry is no longer that high. It is a technique that allows people to walk away with a physical 3D object without having to really do anything. We don't regulate 3D printer machines, so students can learn and explore through it. The more barriers that an institution puts in the way of students, the less prone they are to innovating. It also means that students will usually find whatever is more conveniently available anyway. Typically, the first thing students do after they receive training with the 3D printer is print out stupid little hearts and keychains. But after a certain point, that doesn't matter, and they are able to move on to more important things. 3D printing has been a gateway drug to get more students designing and creating innovative work.

humble beginnings 1 of 7 former PlayLab studios, Brooklyn, NY

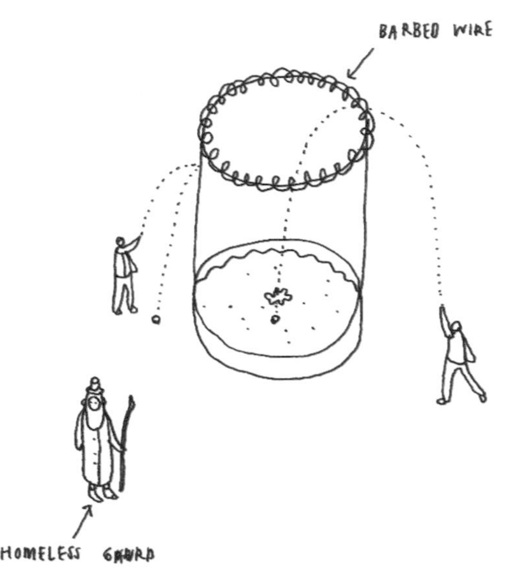

The Secure Well concept drawing *all images courtesy of PlayLab*

What was the genesis of PlayLab?

PlayLab started back in architecture school at Virginia Tech. My partner, Jeff Franklin, and I met and had a like-minded view on things. We acknowledged that we did not have a reference for anything. We were learning so much about different types of design and art, and we weren't interested in making just one thing. It is more interesting to pursue whatever we want to do, like in an art practice or the way a musician works. We had no clue how we would do that. We moved to New York in 2007.

Are there set roles or structures?

Generally, yeah, but it all depends. I'm the "meeting guy;" I prepare a lot of presentations. But everyone pushes the same weight with design. There are projects that we generate from the office, like + POOL, a collaboration with Family New York, which is an architectural office downstairs. We work with a lot of practitioners and offices: designers, engineers, and politicians. We don't feel a need to hire in-house for that.

What is the advantage of a small studio?

It just sort of happened. If the studio booms up to be 100 people, that is fine, but we are not an agency. In a way, we are not a design office; we are an ideas office. We're more like an art office. We work with clients, and we do commissions, and that is an important part of our practice, but we put a lot of effort into generating ideas internally. They are not necessarily for the will of somebody else; they are for the will of us. We are young and relatively new, so we keep things light. We are going after things that might not be financially successful right away.

Is there a link between youth and your process?

We are not willing to sacrifice the quality of our ideas or the way they happen. Some projects, again, like + POOL, are going to take a long time. We have been working on + POOL for over four years. We got into publishing and launched a quarterly art publication with a few other people, and publishing in general is not an easy place to make money. We knew that going into it. We just wanted to make a publication. We are following the things that we are interested in, and we are self-critical about that, but at the same time, everything is done under the lens of fun and play. We all could easily be in different offices doing different projects, and we have all done that. It wasn't fun.

Do you have a specific collaborative process?

No. Every project begins with a conversation, or a joke, and we want to see how far that joke can go. We are having the time of

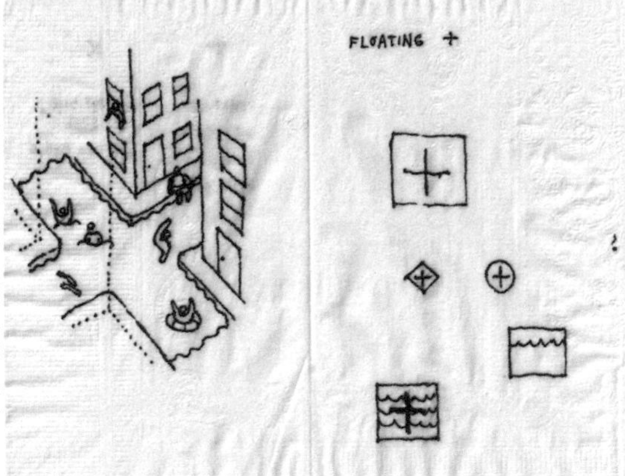

+ POOL water-filtering floating pool for the rivers of New York City, collaboration with Dong Ping Wong of Family New York

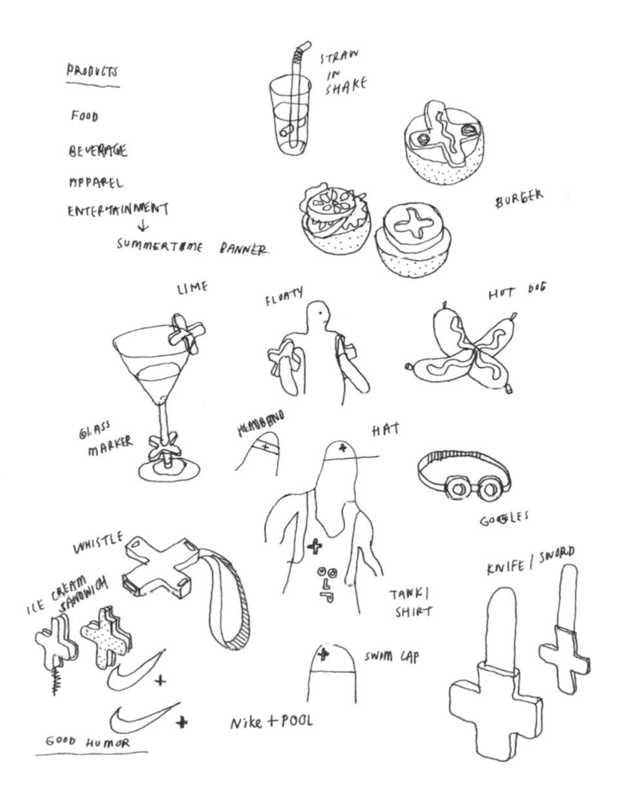

our lives articulating these ideas in a fun way, and chances are, the vast majority of other people are too. And if they don't, we don't care. Maybe that comes from a background in punk. Collaboratively, it depends on the scale. + POOL is the biggest example, because we are working with over two dozen different companies and organizations. There are so many aspects to that project, and we are acting as the managers for the project.

Do you often work with nearby partners?

We are in a building with a variety of people, including the architecture office that we work with, Family. Jeff met them at REX, and then we found ourselves in this amazing neighborhood in West Village, New York. So no, we don't typically work with a lot of people in this neighborhood because, for example, Sarah Jessica Parker lives across the street. In the instance of Family, Dong-Ping had recently started his office, and we had recently launched a project with some friends and designers called Pie Lab, a pie shop in Alabama. John Bielenberg was there for that as well. Pie Lab was a mile marker in terms of the type of projects that we wanted to do, and the way that we wanted to work. Just like the pie shop and just like a record label that we started before that—where we spent two years producing records and meeting bands—we were figuring out how things worked. We've moved around a lot, but now we are in a central part of the city, which is way more convenient than Brooklyn. It's easy to meet out for coffee in the middle of the day and talk about ideas.

Are many interactions unplanned or informal?

Yeah. We are pretty social people. We never wanted PlayLab to be separate from our lives. In this city, the chances of meeting somebody that you can do something great with are pretty high.

You mentioned punk. Is there a music influence?

I grew up in Virginia Beach surrounded by punk. People there were like, "I don't even know how to make music, but let's go try. Let's go buy this shitty instrument and figure it out." So I did that. I played music and I still do. That is really important for PlayLab. We don't ever tell anybody they can't do something.

+ POOL is an incredibly ambitious project. There are political issues, government issues, funding issues, not to mention creating the actual technology and designing the world's first filtration system that cleans the Hudson River passively, so that human beings can swim in it safely. It's all on our shoulders, and from that mindset of, "Fuck it, let's go for it." We work hard and smart, and we're good to people.

Eyebeam Art+Technology Center

Roddy Schrock, Director
Brooklyn, NY

Data Viz Marathon 2010 weekend workshop, impact of humanity's footprint on Earth *photo: Ann Liu Alcasabas*

Why is physical space important to collaboration and interdisciplinary work at Eyebeam?

It is easy to put artists into a technical environment, lock them in a room, and claim that it is an art and technology residency. What makes Eyebeam unique is that we spend so much time thinking about the kind of cohort that we are building, and we develop a space that encourages conversations. Everyone works in a shared studio space. We have formal weekly 45-minute group meetings with every resident and fellow, to track progress on work and help solve roadblocks. We also have monthly shop talks, where we require everybody to close their computers and listen to fellow artists talk about problems in their work. Every one of those meetings leads to somebody realizing something about their work.

It is easy to think about collaboration as a virtual experience, such as online through Skype. But having people physically together in the same space, at the same time, seeing the way other people work, regularly chatting in the kitchen, and having access to the kinds of applications and tools that we provide, gives people a chance for growth that I don't see replicated in traditional residency programs.

One good example is James Bridle, who joined us through a joint residency program in London. He had previously been through residencies where he worked mostly in isolation and did a lot of writing. He started collaborating with resident Ingrid Burrington, and they developed a performative talk, a one-on-one conversation that was presented publicly. I saw changes in the way they were thinking about their work, just based on having the collaborative opportunity. The talk was a culmination of joint thinking through similar artistic problems and issues over the course of 14 weeks.

How does Eyebeam facilitate a sense of play?

Eyebeam has always been a place where people can challenge one another and do things that don't make sense. It is a space that allows people to try ideas and fail. The crazy warehouse space that we have in Chelsea allows for that. You don't have to worry about messing things up. You don't have to worry about whether the space will survive whatever project that you're doing. We are working with the architects to keep that sense of play in our new Brooklyn location. We want a space that still allows for experimentation, but you don't find that a lot anymore. We've been around long enough that we can point to that kind of space leading to real successes. For instance, Graffiti Research Lab met here to develop the EyeWriter. We know that these kinds of projects have been quite successful, and it is because of the kind of space that we offer.

What is the Eyebeam take on tech and culture?

More than ever, technology is being integrated into people's personal lives. Eyebeam has been focusing on issues of digital intimacy and information ownership. Our culture lacks an understanding of the technology that we use. It's a major issue that's not going away. The more that we can bring artists in to explore some of these problems, the better we can prompt the general discourse around these issues. I think art can be a way of helping people digest what it is that they are facing. Artists will always be the ones who are leading creative inquiry. There has never been a time when that isn't the case.

Learning to trust artists is the hardest hurdle for institutions. Artists are great producers, and often work at a level of efficiency and production that is higher than expected. It is hard for companies to understand that, and especially hard for educational institutions to understand that.

Flux Factory

Nat Roe, Executive Director
Long Island City, NY

rafting on the canal in Aarhus, DK *courtesy of Flux Factory*

Flux Thursday monthly potluck dinner *courtesy of Flux Factory*

How is Flux Factory structured for collaboration?

Flux Factory is highly multidisciplinary. The resident group here at Flux Factory is the heart of who we are and what we do. The residency program gives studios to artists who share the building. We also do many classes and workshops. The Exhibitions Program includes four major exhibitions a year. Our residents put on many solo and collaborative exhibitions beyond that, and special events. The exhibitions are usually curated by the residents, and the artwork in the exhibition primarily comes from residents. Flux Factory functions because there is a lot of resource sharing, including meals. If you put work into the collective, and if you are committed to the collective, then you have equal ownership over the curatorial direction of the space. You get the space for the night to put on a concert or a show or a film.

Collectives result in different kinds of curatorial output than what you would see otherwise. If you were to look at a gallery with one curator, there is a cohesion of vision. To a certain level, you give up your own tastes in a collective. You end up with exhibitions, outputs, and artwork that are rich and diverse. On the other hand, you have to live with a lot of imperfections.

One of our recent, collectively produced, exhibitions was *The Exquisite Contraption*, which was a Rube Goldberg machine set in Flux Factory. It's a great analogy for collectives in general because different artists make work that feeds directly into each other, and they are inherently dependent on the cohesion of linked works. The ironic symbolism was that parts of the Rube Goldberg sometimes didn't work, so we secretly had to help it along. Artistic collectives are a microcosm of society at large.

How does Flux Factory play within the community?

Flux Factory definitely does have in its DNA the punk warehouse vibe. Our peers are also galleries and museums. But the true collaborative aspect of what we do comes from a tradition of artists living and working in warehouses. New York City used to be more of an outlaw city. You used to be able to get away with living in a warehouse; it was something that people in the Lower East Side did all of the time. But, there is a genuine, new paradigm in New York of artist collectives learning the lexicon of real estate and aiming for long-term stability. In the past, a lot of people didn't feel like it was that big of a risk. I like to think of Flux Factory as leading that shift for artist-run collectives.

Aziz+Cucher

Anthony Aziz & Sammy Cucher, Faculty, Parsons
New York, NY

Interiors collisions of human and environmental elements

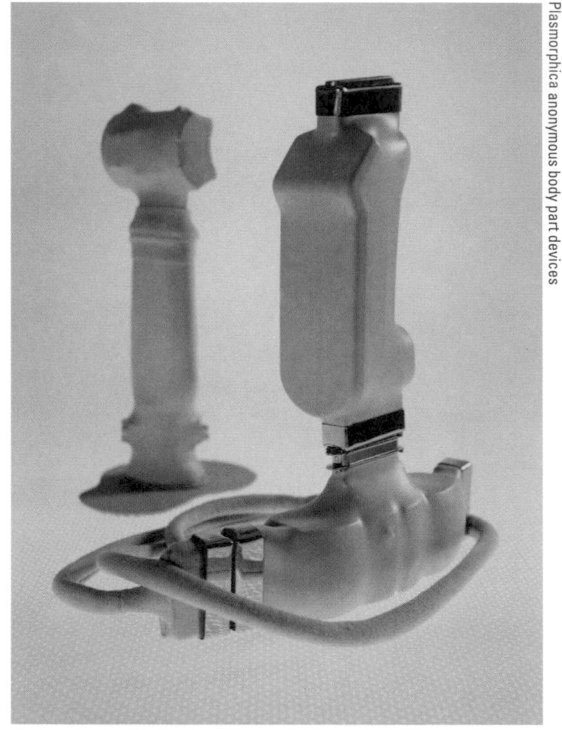

Plasmorphica anonymous body part devices

How do shared experiences impact collaboration?

Everything we do is determined by our day-to-day interactions and reactions to the outside world. Sometimes things are very specific, like all the works from our exhibition *Some People* and the *Aporia Series*, where we explored our own complicated, affective relationship to the Middle East and its geopolitical realities. In 2009, we took an extensive research trip to Israel, Lebanon, and countries in the former Yugoslavia. We came in close contact with the scars of recent and ongoing wars, and that was a catalyst in the production of *Some People*.

In the beginning of our collaboration, we were emotionally marked by the reality of AIDS in our community and in our personal lives. All of our work relating to the body, such as the *Plasmorphica* series and *Interiors*, can be understood within that mind-frame of disease, and control or sublimation of the corporeal, even though it doesn't deal with AIDS in a direct way. We were going to raves in the early 2000s. The lights and buzz of the night life inspired *Synaptic Bliss*, where we played with consciousness and overt digital-ness. There's a collision of perspectives and scales. We were influenced by blending the boundaries of body and environment, as experienced in raves.

Aporia Series #4 politicized clown costumes *all images courtesy of Aziz+Cucher*

Mike Mandel + Chantal Zakari

The State of Ata representing political power in Turkey *courtesy of Mandel + Zakari*

What is your collaborative process?

Chantal: Our collaborative work is all about discussion and analysis. We collaborate on every level of the project and the production is a shared load. It depends a little bit on skills, but it also depends on who has time to do each piece.

Mike: Chantal comes from a design background, and I am a photographer. But we consider ourselves partners in every aspect of the project. Our most recent project, *Shelter-in-Plates*, is a series of commemorative plates designed as a response to the military lockdown in Watertown, where we live, during the manhunt for Dzhokhar Tsarnaev; we were both involved in the concept and design of this project.

Chantal: There were SWAT teams everywhere: in our streets, looking in our backyards, peeking under porches. People were escorted out of their homes because we live in an area with a lot of Middle Eastern immigrant families.

Mike: It was basically martial law without martial law being declared. So we collected all of this imagery: video stills and thousands of bits of information about this event from the Internet. We had already been thinking for the past few months that we were going to do something about Watertown's military legacy, because Watertown has one of the first national arsenals. But without even having to do any research, the imagery came to our house on April 19, 2013.

Chantal: All of this imagery that we had collected about the manhunt in our neighborhood was of houses that we knew and recognized. The idea that we came up with was to design commemorative plates. We found a company in Colorado to fabricate the series of six plates, and we designed them with images that we found online, then coupled it with Victorian floral embellishments. We were able to produce them cheaply enough so that people who may not typically collect art were able to buy them. Now our plates are hanging in dining rooms in Watertown and throughout the country.

What is the role of collaboration in activist work?

Mike: We collaborated on *The Election Campaign*, when I ran as a write-in candidate for Town Council to engage our community about the issues surrounding a Walmart store that was trying to get approval to come into Watertown. We got hundreds of people to help our campaign, and Chantal was the manager.

Chantal: Even after the election, we continued the work by raising money for a billboard we designed, which we installed right where the proposed Walmart was to be built.

Mike: Walmart withdrew their application after our billboard was up for two months. And I almost won the election; we lost by 89 votes.

Chantal: Previously, we worked on a project for 14 years—*The State of Ata*—which resulted in a book and a series of exhibitions. Beginning in 1997, we went to Turkey annually to photograph in the streets. The focus of the project was the symbolism of Mustafa Kemal Ataturk's [founder of modern Turkey] imagery as displayed in public spaces. Ataturk represents a secular and modernized Turkey. This imagery can be a liberating image, but for the religious, it can be seen as a more oppressive image of limited freedom. We documented his image, interviewed people in the streets, and invited them to pose in front of his photograph in our popup photo studio in the middle of the street. There was an Islamist protest march, and I stood there with a little image of Ataturk, and Mike took a picture. That photo ended up on the front pages of the national newspapers, labeling me "The Girl of The Republic," and became a publicly recognized symbol of the current political conflict in itself.

Aaron Gotwalt

Starter of Startups CoTweet, Seesaw, & Everlapse
San Francisco, CA

CoTweet original logo courtesy of Aaron Gotwalt

Can you riff on your background with startups?

I left my job at an ad agency in 2008 and moved home with my parents in Lancaster, Pennsylvania to do consulting and pay the bills. I had this crazy idea that I was gonna build this stupid fashion app on Facebook—it was going to <reverb> change the world </reverb>. In the process of building it, I worked with a friend of mine, Kyle, who was also still in Lancaster at the time. We first started working together around 2002. Kyle is an extraordinarily intuitive designer. For him, design starts with pencil sketches and translates into exquisitely crafted HTML and CSS. If you ask him to theorize his way out of the bag, he can't. If you ask him why he's doing things, sometimes he can't explain it. But he's really good, and we've been together awhile.

The Twitter thing was happening at the time, and we were using it to promote our fashion app. In doing so, we figured out that there were some real basic problems with using Twitter to represent a product or team. So we decided to build what I describe as a "palate cleanser." We spent three weeks building something totally unrelated, curious about what would happen, and it became CoTweet. There were companies who were trying to buy access, but we didn't know how to talk to them or sell them stuff. That's when my friend Jesse, who works that angle well, offered to help. We started CoTweet in November 2008.

We hired an engineering team and built a really solid product. Huge brands like Ford, Microsoft, JetBlue, Coca Cola and others all used it. We quickly found ourselves needing more money to keep building and growing, and deciding between raising another round of financing or selling. We accepted a flattering offer from ExactTarget, perhaps the best decision of our collective lives.

It was a strange transition though—our fledgling seven-person team quickly became part of a 600-person company. For all their best intentions, they would say things like, "Hey, we need this to be in Japanese." So then everything that we were doing would stop, and we would spend two months trying to convert the thing to Japanese. They had no idea of the level of complexity to do that, vs. to do something interesting.

The product stopped moving forward at the exact point we sold it. It's interesting how much we accomplished in less than a year leading up to the sale, and how little we accomplished in the following 16 months, even when we had many more resources to play with, and despite our best efforts. There's a hustle or drive when it's your thing, and you're trying to make something. I had the best and the worst part of being obsessed with my product. I came out to San Francisco, worked 80 or 90 hours a week, had a really nice apartment that I never really saw, knew nobody, and felt really isolated. Selling the company for what was a truly life-changing amount of money was not satisfying. I was manic. The company was something that I ate, breathed, and slept. Few others in the company were like this, except Kyle and Jesse. It was really hard to get anyone else to think of it as more than a day job.

Kyle moved back to PA and bought a castle in Lancaster. We're now doing the remote thing, and he flies out once or twice a month.

Then, Jesse called with an idea that became Seesaw, a collaborative decision-making application. I don't care what 14-year-olds think is cooler: a Ferrari or Lamborghini. It just doesn't matter all that much. What does matter is when you're in the shampoo aisle, and you're overwhelmed by choice. If I can build something that will always help a consumer pick one brand over another, then I'm actually building an interesting business. Launching was easy, but we quickly found out that it was stuck somewhere between a utility and entertainment. You don't need structured data and 400 opinions. People used it more when they were looking for affirmation of decisions that they already made: "I just bought this shirt, isn't it cool?" You always want to answer, "No," but people are surprisingly positive in that kind of scenario.

Our investors were super supportive despite the result. Kyle was sitting back in Lancaster at the time when he had this alternate idea for a product called Everlapse. It was like a collaborative flip book, but with some cool onion skin. The thought was that it would make the barrier to participation really low—it allowed you to just go and do something. It didn't do poorly, but I think it made me realize that there are so many things competing for peoples' attention.

When CoTweet happened, Kyle and I initially built it,

but Jesse came in super early to run it, and that created an interesting set of internal politics. So we tried to align things a little bit better in the Seesaw experience. We worked really hard at it, but I think that there was tension in the team and the shuffling of roles. It didn't help that Kyle was working remotely. One downside about remote collaboration, I found, is that the subtle communication and dynamics of in-person communication get lost, even with video calls every day. For instance, sometimes you're left wondering if this person is disagreeing with you, or, are they a bad human being? You know this person has never been evil to you in the past 10 years, but on the other hand, maybe they're starting now.

There was one moment at CoTweet where Kyle had just flown in, and we were arguing about something. We walked from my house for a mile, screaming at each other, and it wasn't even good rhetoric. We are fine now, but because Kyle and I have worked together for so long, our professional relationship is sort of frozen in an emotional maturity of 10 years ago. That can be hard for the other people in the office. Kyle and I figured out that we'll have these terrible arguments, but we'll be fine; we're both on the same team, trying to solve the same problem, trying to get the correct answer, and we're willing to be wrong.

We actually hired a business coach to help mediate the tension that built up. He had all these frameworks and hand-waving ideas that we kind of laughed at, but he really helped. It's amazing how your own perspective can differ from the reality.

Now I'm starting something with a totally new team. I'm having fun working with people that I've known really well, but haven't worked with professionally. I'm probably going to continue working with Kyle, just maybe not on my foreground projects. We work really well together. I tried to work on something without him one time, and it was like not having an arm. But I also think there's something really healthy about a little professional distance, where you don't have this body of experience, both positive and negative, to draw upon daily.

How do you choose your team?

A good startup team is a function of time and money. Our high point at Seesaw was eight people. At Byliner, we were roughly 30. The reason why I brought people from the East Coast, and people that I've known for a long time, was really just a supply and demand issue. There is so much demand at an early stage, and it's very hard to find qualified engineers to do things; often times it's just easier to pull from your social network.

I have a bunch of ad agency friends, IDEO-type people. I think IDEO represents an alternative to academia, but it's not quite the real world. They make you design things that are interesting but totally not commercial—like in a world where you make tea without using water. It's interesting to watch people like that try to make the jump, because it's very trendy to be in startups right now. The IDEO kids try to jump into something that is extremely practical, and less comfortable. You're a senior designer, and you're cleaning the toilets; it's a weird mix of responsibilities. These folks go from changing clients every few weeks, to investing their soul in doing something that may or may not work.

What is the division of labor and its impact?

I qualify people as either "product people" or "not-product people." Meaning, there are engineers who don't care what they're working on; they just want it to go extremely fast and be algorithmically challenging. They get lost in the obscure details, but they don't consider the larger context. You need them as the company gets bigger, because you need worker bees. Worker bees are effective, they get things done, and they don't often ask questions.

On the other hand, you're not sure what you're building early on, and if you don't have a lot of people asking constructive questions, then that's a real challenge. How do you bring together a team of under ten people, where most of them have opinions about the product and business that you're building? How do you build an authority structure that isn't entirely destructive? At that stage, what you're building could turn into anything. Your purest true believers, your product visionaries, get disillusioned with the product as it becomes more about optimizing and making the ads work. As you grow, you need middle managers to deal with scale, and you need drones who do stupid things because they're told to do so.

Does a physical working environment matter?

You need people to be face-to-face all the time. Sometimes you need to take a sharp left turn that you weren't prepared for. If we go out to lunch having this funny conversation, and then there's this idea moment that changes everything about how we're going to operate, the remote person will have missed that. It's hard to get long-distance people to work effectively together, despite some great software. But at that early creative stage, it doesn't work well. The best moments at Seesaw were when Kyle was in town, and we would have all these great ideas. If we had more of that, I think we could've gone further.

Free Range Studios

Louis Fox, Co-Founder
Oakland, CA

How did you and Jonah Sachs meet?

We were friends as kids in Woodstock, New York. We met in a private school called the Woodstock Children's Center. This was in the '80s, so most of our teachers had been at the original '69 Woodstock Festival as teenagers. Jonah and I had a pretty progressive childhood, and that definitely influenced us later on. I remember watching the movie *WarGames* with Jonah. It was about the potential for nuclear disaster, and we were really inspired by that. We started coming up with a bunch of PSA ideas. As kids, we figured out what was right and what was wrong together. Our political beliefs are basically identical, and we still try to make sense of things together via our work.

How did that friendship become a business?

Jonah went to Wesleyan University and I went to SUNY Purchase. Jonah majored in American Studies and became Editor of the school newspaper. There, he picked up some print design skills. I went to film school. After school, Jonah moved to DC and started freelancing. When I left film school, I became a Production Assistant for low-budget, independent films in New York City. I worked on a bunch of commercials. No surprise, I didn't like doing this very much and had no interest in climbing the production ladder. In 1999, I moved to DC and lived on Jonah's couch for several months. We built our business by finding lots of freelance and spec work.

Can you describe a collaborative project?

In 2003, Jonah and I wanted to boost our studio's reputation. We knew we wanted to make a Flash movie, because people were switching from video to animation. We got a grant, and we put out the word that Free Range was going to give away one of their Flash movies. We put up a request for proposals, and a few of them centered on the factory farming issue. Then we had the idea to combine three clients into one project. We worked with the Global Resource Action Center for the Environment (GRACE), Farm Sanctuary, and People for Ethical Treatment of Animals (PETA). The politics between them were different flavors, but Jonah and I saw the overlap and were able to work with the politics amongst these different groups. We wanted to design something that could be used by all of them, so we ended up making *The Meatrix,* and it was a success. Forcing these groups to collaborate enabled the animation to go viral. To solve for variances in opinion, we gave it different endings.

Louis Fox (L) + Jonah Sachs (R) produce documentary on eccentric Las Vegas residents, origin of FR Films

Grocery Store Wars organic foods movie *all images courtesy of Free Range Studios*

Design RePublic

Lindsay Kinkade, Founder
Phoenix, AZ

Phoenix Design Week Lindsay+Ruben Gonzales lead street stencil-making workshop *courtesy of Design RePublic*

What are the origins of Design RePublic?

I help people understand what Design for Social Change is. My background and first career was in journalism, working at *The Boston Globe* for seven years. I was a newspaper designer. I loved my job and the people that I worked with, and I loved storytelling and writing; however, I did not like sitting at a desk. I felt that every morning, the reporters and photographers got to go out into the world, capture what was going on, and talk to real people on the street. But, so much about newspapers was changing. My roots are in journalism. The new economy, crossed with a studio and a backpack, equals a new, small, and agile studio model.

How does the space allow cross-discipline work?

I first did an R&D experiment to figure out the studio space I would need in Phoenix. I built the studio furniture as a co-lab effort. I determined what we needed to change the room to facilitate various activities: tools for small meetings, vs. bigger meetings, vs. work sessions, vs. intense design-athons. We built furniture out of plywood and sawhorses, but it all folds flat and fits in my tiny station wagon. It had to be lighter, faster, and cheaper, so that any of us could take it to where we needed it to go. I found the space later—the studio is a 1000 ft^2 warehouse in a neighborhood that artists colonized 20 years ago. The space allows us to make a big mess. I bought an Airstream, so now we can take it mobile. Finding the right landlord is a big deal. Our landlord is an artist; she lets us do things that other landlords wouldn't let us do.

How did you reach out to the community?

It takes a long time to build trust and do this kind of work in the community. I had to do a big social networking course to figure out who was here, what they were doing, and how my skills could fill some of the gaps without stepping on anybody's toes. My collaborations are built on a period of deep immersion, going to community meetings, festivals, other people's events, gallery shows, and putting on my grubby shorts and pitching in.

I've bought into the idea of the New American University, so I wanted to be near Arizona State. We have thousands of people graduating every year who believe that the world can be a better place. But, we still have a lot of work to do in building specific pathways. In a commercial economy without a lot of social practice, it took a long time for me to explain what Design for Social Change even means.

Detroit SOUP

Amy Kaherl, Director
Detroit, MI

Detroit SOUP November 2013 photo: *Dave Lewinski, courtesy of Detroit SOUP*

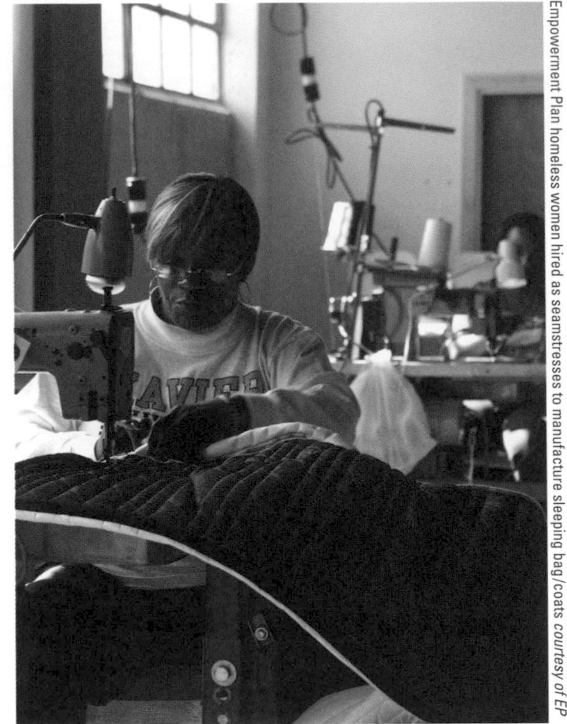
Empowerment Plan homeless women hired as seamstresses to manufacture sleeping bag/coats *courtesy of EP*

How did Detroit SOUP start?

It was founded in the Mexicantown neighborhood in 2010. Kate Daughdrill experienced a similar dinner in Chicago with InCUBATE, who originated the SOUP idea of collaboratively giving micro-grants through a meal event. We wanted to collaborate on an idea that could empower and connect the community while exploring new art practices.

How has Detroit SOUP grown?

We are now a project of Cityscape Detroit, a 501(c)3, and have expanded to empower residents of various neighborhoods. Each neighborhood has a collective board that shares duties and responsibilities of gathering locals together to find new ways to make their area better. We continue to run a monthly citywide SOUP, which brings new folks to collaborate.

What are some notable project examples?

In 2010, a college student designed a coat that would turn into a sleeping bag for use by the homeless. She hired women in shelters to make the coats, and she now employs about 20 people for the organization: The Empowerment Plan. The model inspired a couple of jewelry designers to hire women from shelters to make jewelry from graffiti paint chips that have fallen off walls. Their business is called Rebel Nell.

There is also a poetry group that formed to present at the Livernois Corridor SOUP. And, Obsidian Blues Detroit uses art to transform, heal, and forge new ground.

How is the gathering integral to arts collaboration?

It is an important tool to collaborate, not just within the arts, but across all ideas that interact with social innovation. Each one of us has a small network of people that can help, inspire, or connect people with social or physical capital. It is amazing how this is turning into a new town hall gathering, where people can find a safe space and see how people are creatively thinking about the work that is happening in their city.

A simple space allows us to hear these concise ideas. The space does not need to look like a wedding reception. Expectations change when there isn't a lot of glamour. You can hear the ideas, and you can see the people. Soup and salad make it easier for people to gather and have a conversation from a shared experience. It's a very human experience: eating together, sharing in storytelling, and sharing values and beliefs by just asking the question: "What project would you, as a citizen, support?"

Evan Roth, Co-Founder
New York, NY

Laser Tag on Brooklyn Bridge, GRL + Agent Watson + Bennett4Senate + Huib Van Der Werf *courtesy of Evan Roth*

We saw ourselves more as Q Branch than James Bond. We would be the person who supplies the tools, rather than the person who uses them. Our idea was that we wanted to make tools that would level the playing field between the kind of scale that cities and advertisers could take on, vs. the people who live in the city. The other goal was to expand the sphere of our open source work.

Our first project, *LED Throwies,* ended up being a big hit for us in terms of how it traveled around the web. We would have a pop media piece or video that was meant to be passed around blogs, and we also had source code, or how-to guides, that we paired with that media.

The one thing we realized was that the technology as a meta-project was more of a cultural hack than anything physical or digital. We realized that the projects that we were doing were getting reported on by the media disproportionately to what they should have been. The media wants to write the story that *We Are Living in the Future.*...We quickly figured out that if you just sprinkle lasers and LEDs on stuff, the chances of getting your political message spread are a lot greater. The tools that we were making didn't matter, as long as they shined enough to show up in a nice, pretty photo.

Can you describe your time at Eyebeam's OpenLab?

Unlike any artist residency or fellowship that I have been a part of, they were looking for open-ended research. They didn't treat art like a deliverable. They treated it like a research process, which I am really grateful for, and has influenced me to move past deliverables. The only stipulation was that the work had to be totally open. If it was media, you could use Creative Commons. If it was a hardware project, you could release a how-to guide. When you have a whole lab full of people that are all working in that model, the ownership of the work becomes irrelevant. I worked a lot with friend and collaborator, James Powderly. We ended up not even putting our names on anything that we did together. One person threw out an idea, and then another person built back on top of the other person's documents and comments.

Can you talk about Graffiti Research Lab?

I presented some of my graffiti research that I was doing and some of the software that I was working on. James had worked on several robotics projects and volunteered his skill set and experience. He provided materials and hardware expertise, and I provided software and access to the graffiti community.

Has the nature of your collaboration changed?

I don't try to earn a living on the collaborative work. When I work on collaborative pieces, it's because I really want to, because it's fun. Collaborative models that fund food and rent are really difficult. It's the Wu Tang syndrome. You can't have 11 dudes in a group, go on tour, everybody comes home happy with each other, and everyone is rich and famous. There is already such little money in the arts, so dividing a small pot into even smaller sub-pots just doesn't work.

The more you can remove money from the creative process, the better. Everyone is broke as shit in the arts, scraping for meager amounts of money that trickle down from the government, companies, or individuals; all of which are underfunded. When you throw a few breadcrumbs at a bunch of hungry fish, of course it is going to cause ripples.

Is having a solo outlet essential?

It is good for me to have an outlet where I don't have to make concessions. To be respectful of the people you are collaborating with, you never publish without consensus.

Jessica Hische + Russ Maschmeyer

Don't Fear The Internet & Letter Together
San Francisco, CA

Letter Together weekend workshop, participants collaborate on alphabet *all images courtesy of Title Case*

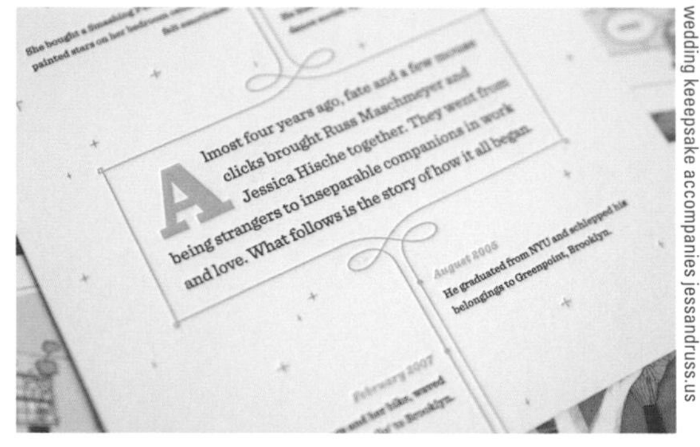

wedding keepsake accompanies jessandruss.us

tutorials for non-web designers by Jessica + Russ

What is your individual vs. collaborative work?

Jess: We do very different solo work. I do lettering, illustration, and type design. I work for myself, which means that I am endlessly doing freelance, client projects, and self-authored projects, which is very different from how Russ works. Russ is extremely collaborative; he has to talk to people all of the time.

Russ: I do product design for Facebook. I work on searches within Facebook. It is firmly rooted in technology design.

Jess: One of our collaborative projects is *Don't Fear the Internet*, which is a website that Russ and I created together. Well, I made it, then forced Russ to work on it with me. Our relationship excited me to try the web, because living with someone that does it every day allowed me to pick his brain. I wanted to share my newfound joy with novices, while Russ could give more technical advice. Russ was a graphic designer before grad school, so we have that shared world.

Russ: We have focused more on interaction over time.

Jess: Russ was also in a band while we were together. He told me that the main reason why he got into graphic design was to make stuff for the band, which I think is true of many designers' origin stories.

How do you resolve varying preferences in shared working spaces?

Jess: I have been in situations with studio mates where there were conflicting ideas about what the environment should be. Erik and I both like an active, talky environment. We play loud music and shout to each other, and it is not an interruption. But when there are many people, and one person requires total silence, everybody else is in headphone-aggravation world. You can establish quiet hours, so that everyone can answer emails, but then in the afternoon, it is loud and fun.

If you switched jobs for a day, could you hack it?

Russ: What Jessica does is client management. She works directly with her clients. In-house is different. I have established trust and deeper relationships with co-workers on a longer-term basis. But Jess always has to be in that customer service mode, which kills me. That would actually be the hardest part of Jessica's job for me. Spiritually, that would be so draining.

Jess: Russ is a manager right now. While he's a good product designer and thinker, he struggles with cheerleading and people-wrangling. I would kill it at that part, but I'd have terrible ideas about Facebook.

Anna Wolf + Mike Perry

Anna Wolf Photography & Mike Perry Studio
Brooklyn, NY

Layers of Color *courtesy of Mike Perry Studio*

Anna on set managing photo shoot *courtesy of Anna Wolf Photography*

How do your workspaces relate to your work?

Mike: When I moved to New York, we started working next to each other in the same home space. But, we outgrew the apartment, and I desperately needed a studio. We both moved into the studio, but after about four years, Anna realized it wasn't the right working environment for her and moved back home. I've taken over the entire studio, and now it's a shitshow. Separate spaces are really important to us. She loves the luxury of going downstairs and making a salad. Anna's office represents her: quiet and peaceful. My space is the opposite: colorful and loud. The music always plays too high.

Anna: When Mike is in his studio, he's making things, he's being creative, and so the music should be loud. When I'm in my office, I'm working on treatments for clients, invoices, or I'm talking on the phone, and doing email—the kind of work that deals with the minutia of production. My other office is the world; I shoot on location, so I'm always traveling to other cities. Part of the reason why working in the studio didn't work, was that I'm always gone, and it felt like three different homes.

What is your collaborative process?

Anna: Mike sees the bigger picture. I have a more practical mind, because that's my job—I arrange photo shoots for a living. Mike has all of these amazing ideas, and I'm always like, "Yeah, that sounds fucking amazing, but how are we going to get that done?" Photography is no joke. It costs a lot of money and involves a massive group of people. Mike is very prolific, but when it comes to big productions, I am the one in the trenches.

Mike: Our most recent commercial collaboration was easy because I had creative control, and then my role was to say, "Anna Wolf, I want you to make these pictures magical."

What project sums up your collaboration?

Anna: We are working on a magazine together, called *Tidal*. It's our baby. It all started last summer, when I was having a problem getting a fashion story published. Mike suggested that I make a magazine for myself, that way I could take control over that aspect of my work. Working with Mike for a while, I feel like I am now in a place where I want to talk less, and make more.

Mike: Yeah, we have three-hour-long, epic walks on Saturday morning for work-strategy conversations. Those conversations are incredible, because we're both equally invested. It's very pure. You know that the feedback is coming from a very trusting place.

0 TO 1

Garner Oh & Tamara Petrovic, Co-Founders
New York, NY

Con Artist Collective main floor of the makerspace *photo: Kanako Miyamoto, courtesy of 0 to 1*

How do your individual skill sets and backgrounds impact collaborative work?

Our scales are very different. Garner is an architect and Tamara is a furniture designer. We help each other zoom in and out.

What is the 0 to 1 design process?

Projects initiate through conversations with the client, whereby we consider the objective, budget, location, timeline, aspirations, and other details provided by the client. A concept emerges from these parameters. During the creative process, Garner and I spontaneously take turns developing the design. Usually our ideas run in the same direction, so it feels like one person is thinking. Sometimes one of us changes directions, then the project follows the lead of the other person. Garner is more practical and technically oriented. Tamara enjoys experimenting. The client's circumstances guide the life cycle of the collaboration.

Do you think it is important to focus on collaborative work or individual projects?

If a project requires expertise in multiple disciplines, then there is more collaboration. As important as it is to collaborate, it is also important to develop as an individual designer. This can take many forms. The client is also a collaborator. A good client can make a fantastic project possible.

Can you describe the creation of a collaborative piece that overtly represents your individual perspectives merging into one work?

The Re-mind project started with Tamara's interest in exploring vanishing. She was interested in creating a structure using a material that would disappear. However, Garner explored the idea in a completely different way. By prioritizing the ease of production while keeping a small budget in mind, we ended up with a very basic and poetic solution.

Do you have any advice for students working in interdisciplinary collaboratives?

Respect your own strengths and weaknesses and those of others. Having the grace to accept things as they are, is handy.

Does your studio space impact collaboration?

Working in shared workspaces with other creative people works well for us. As a small growing firm, 0 to 1 offers the flexibility that we need, as well as an opportunity to interact and exchange knowledge and ideas with people in other fields.

Toormix Atelier

Ferran Mitjans & Oriol Armengou, Co-Founders
Barcelona, Spain

Toormix Atelier shapeshifting annex space used for co-ideation *all images courtesy of Toormix*

calligraphy workshop cor ducted in Tocrmix Atelier

How does the new space encourage interaction?

The important thing about the space is the format. It must be relaxing and comfortable, but different from the everyday format of the office. In other words: chairs out, meeting table out, computers out. It must be open, informal, customizable, and like a living room. It also helps that it is located at street level.

We wanted to create two distinct workspaces. One to establish dialogue in an informal way (sofa area and swings) and one suited for work and craft (central large table with stools). The tables fold as slates, which is useful as a workshop space. We also try to facilitate interactions between the areas to make it as comfortable as possible.

How is the space specific to Toormix?

We are a small team, so the architects—vora—designed for our scale: a 20 m² space equipped for multi-shape formats of working. It is roomy enough to not feel closed off, but open to make it easy to talk, think, and work. The space is built with lightweight, inexpensive, and mobile materials. The white color feels more open and the wood feels warm. Objects like cups, plates, and pencils are all yellow to match the corporate color of our studio. The purpose of the atelier is for creating sketches, ideas, and prototypes, rather than finished solutions. It is an intermediate step between ideas and detailed work that we develop later in the studio with the right equipment.

What activities are ascribed to the Atelier?

For one, we meet with our clients to facilitate new creative dynamics, similar to a Toormix design-thinking exercise. Also, the space gives us a perfect excuse to invite other professionals to come work together, or meet us personally, which is a good way to generate possible future collaborations. This last activity is in an "aperitifs" format, to promote easy and casual connection.

Not being the "official" Toormix office, this space makes it much easier for visitors to be more open, more creative, and less focused on the usual business. It allows opportunities for collaboration beyond standard interaction, opening the range of possibilities of working with clients and visiting professionals.

The space was built to facilitate inspiration and thinking outside of our office, which is only ten minutes away. We work in the office four days a week, and spend Fridays in the creative dynamics of the Atelier, often exploring self-initiated projects.

Quest University Canada

David Helfand, President & Vice-Chancellor
Squamish, BC

Breakout Room *courtesy of Quest University Canada*

How does Quest University utilize collaboration to facilitate an alternative approach to education?

The most important symbolic thing is the academic building—there are no edges—it's a circle. The classrooms are all arranged with 21 chairs around a table. We never have more than 20 students in a class, and the average class size is 15. Most classrooms are set up with the tables configured in this oval-shape where everybody sits around, but in two minutes, you can quickly reconfigure the room. Across from every classroom, in the center part of the circular building, are "breakout rooms," which include four chairs, a table, and a whiteboard. These are intentionally tiny rooms. For instance, I could spend the first 20 minutes of class setting up a problem, then count off students into groups of four and have them take 45 minutes to frame their side of a debate, or interpret a piece of literature, or workshop their papers.

We assign offices to faculty by lottery, so we end up with a musician next to a mathematician next to an economist next to a poet next to a philosopher next to a physicist. This idea of promoting forced connections was important to us, because it confronts the notion that all academics with PhDs go through the process of learning more and more about less and less, until they know everything about nothing, and then stashing all the people who know everything about the same nothing in the same corridor, in the building behind a secret locked door. Faculty can actually learn new things by talking to the person in the office next to them. This has led to things like our innumerate music professor and our tone-deaf mathematician having offices adjacent to each other, then co-teaching a course on the mathematics of music. This models collaborative learning for students, because faculty are actually learning from each other in front of the students.

It's critical to note that we teach on the block system, which means that students take one course at a time for a month. Most of our courses are disciplinary. However, the first course all students take as first-years is called the Cornerstone Course. That course is deliberately highly multidisciplinary, in order to get students used to the idea that they can't just think one way. By having deeply studied philosophy, mathematics, physics, biochemistry, and economics, they understand that the problems with the world's freshwater supply, globalization, and climate change are not going to be solved by people who are narrowly trained in one discipline. They are going to be solved by people who can move across disciplines. Collaboration is a key part of education at Quest, because all of these kids are coming from high school, or transferring from another university, and the ubiquitous model is competition. I love Ken Robinson's description: Consider when a few people get together, work really hard on a difficult problem, and come up with a really creative solution. In a university, we call it cheating. In life, we call it collaboration. Collaboration is highly valued by a lot of organizations. At Quest, we don't consider collaboration as cheating. In fact, the majority of assignments in every class either pairs students up, or puts them in groups of three or four. Whether it's tackling a set of problems or preparing presentations, students are working together.

Does collaboration preclude peer competition?

There are two aspects of competition. There are internal competitions. Say you set up four groups, and everyone in a group relies on each other. The dynamic becomes one group against the other three groups, or us against them, because every group wants their group to be the best.

The other kind of competition is grades, and Quest does as much as it can to discourage that. It still exists, unfortunately, because a large fraction of the rest of the world still works that way, and pre-law students or pre-medical students need grades to apply for certain programs at other schools. When we started Quest, we really didn't want to give grades or focus anything around grades. But we determined that, because we had no faculty ranks, tenure, departments, or majors, that we really had to do something to make the world take us seriously. We couldn't come off as a completely flakey school. But we made a stupid mistake by not thinking about how a large portion of our students depended on scholarships. One of our students suggested a very clever alternative, and it's something that we now follow: Students don't have to maintain any GPA to keep scholarships. But underneath every student's grade every month, there's a checkbox for faculty that says, "This student is gaining from, and contributing to, the intellectual environment in my classroom." Yes or No. Students need to get at least six out of eight checked each year, in order to stay at Quest.

I once spoke to a bunch of fourth-graders at a school in New York; they were incredibly enthusiastic and had questions for over an hour. But when I got back to my students at Columbia College, ready to discuss a similarly amazing neuroscience paper, they were disengaged. It was a seminar, so we were sitting around the table, and I asked, "Why aren't you all more like fourth-graders?" It was a rhetorical question, but being first-year Columbia students and not recognizing the rhetorical question, five of them raised their hands and told me why they were not more like fourth-graders. One kid actually said, "Professor Helfand, you have to understand that, this is a seminar." I responded, "Yeah, I know. The University spends a lot of money making these tiny classes so you can actually talk to each other." And the student said, "But since the point of Columbia is to come out on top, and destroy everybody else, asking questions is a sign of weakness, so you never ask questions in seminars. You only make statements." You see, in a lecture, you can raise your hand and ask a question because the professor has no idea who you are. But in a seminar, professors supposedly know students' names, and students don't want to appear weak in front of their colleagues. The point of Columbia is to beat everybody else. At Quest, that is exactly the kind of competition we are trying to eliminate. That kind of competition is not productive.

Is education narrowing because jobs are specific, or is there still a need for generalists?

Last month, I was in a meeting with The Conference Board of Canada, which is a think tank funded by big corporations. They are doing this huge, multi-year study called, Skills and Post-Secondary Education. But really, it's more like, "What's wrong with post-secondary education in Canada?" And I could tell them exactly what's wrong. But anyway, they had this forum where they invited a panel including provosts and deans. Companies don't want graduates who can write more lines of code per hour, or who know how to better manipulate Excel spreadsheets—which is what the standard government expectation of universities is here, and why universities are failing our students. They want people who can write and speak effectively and persuasively. They say they want people who can collaborate with each other across gaps in age, background, and department, and get problems solved.

The final important factor to companies, which surprised me, was emotional maturity. They are not finding this among graduates. They want people who are respectful when working with people from different backgrounds. And people who see failure as an opportunity to learn, rather than as a disaster. We emphasize that at Quest. In fact, we have a faculty member who makes her students keep a failure log. If they don't have enough failures by Friday, she takes points off their final grade.

Bleu Acier

Erika Greenberg-Schneider, Founder & Master Printer
Tampa, FL

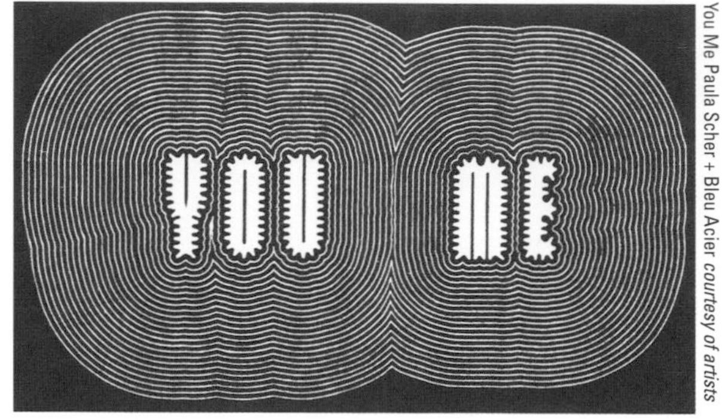

You Me Paula Scher + Bleu Acier *courtesy of artists*

On Off Abbott Miller + Ellen Lupton, photogravure: Erika Greenberg-Schneider *courtesy of Graphicstudio + artists*

Are your relationships with artists collaborative?

It changes from artist to artist. Certain artists have an established history; they have a developed body of work and they know what they are doing in printmaking. In this case, I become a facilitator. This is an example of a silent collaboration.

Then there are the virgin printmakers, and I've specialized in working with them since I started Bleu Acier in 2003. Generally, these artists have never made a print, and have never been interested in printmaking until their work begins to sell. A gallerist or publisher will suggest publishing prints, and the printer has the culture and referential knowledge to advise.

This kind of collaboration ends up working, or not working, within the first hour. The artist either likes or dislikes the perception that the printer has of their work. Starting with their work as a foundation, you keep questioning how it can translate into this other medium. I want it to be their image and not mine. There are moments in this relationship where the artist has to allow me ownership before I can give the image back to them.

How do you decide which artists to work with?

Shops either tend to be fine art print publishers or contract presses. I do both because I need to earn my living in order to print what I want, but I do both lovingly. I also teach because it provides me with the viewpoint of a young set of thinking processes, which helps me grow.

I only work with people on contract projects who understand what my specialty is and the skill sets that I can offer. I learned my trade in France and lived there for 25 years. I love the way French artists think about space, and the French painters and sculptors that I've worked with consider printmaking a high art, and me, a true collaborator. I live and work at Bleu Acier; therefore, an artist who comes to work with me also lives and works in the same space. Artists become intertwined with my family. It is important to me that collaborations go beyond the shop. As a master printer, you are intimately involved with somebody else's work, so the process can't work unless you know a lot about them. A walk together impacts their ability to open up in service of the artwork.

What is your collaborative process in the shop?

Usually, we start by discussing size of the image, process, number of colors, size of the edition, and paper choice.

I worked with Abbott Miller and Ellen Lupton when I was at Graphicstudio, USF Tampa. They were so used to working together, and suddenly they had printers involved. Initially, there was difficulty in comprehending the goals of the image. Their typography research was extensive and influencing the goals for the edition. The first results were good, but not spectacular. Part of the trouble was in the differing ways that we saw space. Type on screen is not the same as type within the materiality of printmaking. Scale in material processes is also far removed from the idea of digital scale. What made that collaboration great was everybody could roll up their sleeves and not have issues with artist/designer relations.

Baltimore Print Studios

Kyle Van Horn & Kim Bentley, Co-Founders
Baltimore, MD

Kim + Kyle in Baltimore Print Studios *courtesy of BPS*

How do you compliment one another?

We used to joke that Kim was a designer with little printing experience, and Kyle was a printer with little design experience, and together we would make up one complete design and printing shop. Since we started the business, we've taught each other about our respective disciplines and continue to learn from each project, client job, and workshop that we do. Kim adds an organized, clean design aesthetic to the printing processes, while Kyle adds his sense of humor, and troubleshooting skills to all of our jobs.

How do you co-manage your print studio?

Kim does more of the digital design work for the studio, which is either for clients or in-house projects. Art direction goes both ways. Kyle manages the maintenance of our machines. Most tasks are a joint effort—from big purchasing, or major shop decisions, to smaller day-to-day management. Both of us are here when the studio is open for rental, because we both help renters, as well as work on personal projects and client jobs, work with our intern, or keep up with organizing and cleaning. We co-teach our workshops and step-in when one of us gets side-tracked.

What piece benefited from dual perspectives?

In 2011, we worked with Radica Textiles, a local studio specializing in hand-drawn and screenprinted textile designs, to create an installation in the entryway of Union Mill. It was an unusual job for us, but we got to push the limits of our design skills, technical abilities, construction, and research.

We also did a fun print with Kyle Durrie of Power and Light Press and Type Truck, an old truck outfitted with a printing press. Kyle drives across the country doing mini-workshops and events. We joked around about how printers and coffee go hand-in-hand and came up with a phrase. She went back to Portland, carved a linocut of a French press, and then shipped it to us. We added type and printed, "The Most Important Press in This Shop." Speaking of coffee and printers, our latest collaboration is with Thread Coffee down the street, a small-batch coffee roaster. We pitched the idea of a custom coffee roast, perfect for our habit of all-day coffee drinking. After many months of taste-testing, the coffee is now available. We completed the project with label designs. The hand-set wood and metal type was scaled down for digitally printed labels, but we also include a free, full-scale print with every order.

Post Typography

Bruce Willen & Nolen Strals, Co-Founders
Baltimore, MD

Mid-Term Election Fold-In *Washington Post*

Blood Brothers Client: Ottobar *all images courtesy of Post Typography*

Bruce (L) + Nolen (R) Post Typography studio

How did you meet?

Bruce: Nolen and I met in art school at MICA around 1999. We were both into similar types of music and went to shows together. The first time we started collaborating was when I came up to Nolen over the summer one year and said, "Hey, do you want to start a heavy metal band?" Since we had a band, we naturally had to make shirts, posters, stickers, etc.

Nolen: We were the only two people I knew of, who collaborated in school. Unlike Bruce, I wasn't a design major. I initially went to school to be an illustrator, but I changed my mind pretty quickly. By my junior year, I was screenprinting posters for shows around Baltimore. Although, I wasn't really thinking about graphic design when I made them.

Bruce: I'd seen Nolen's work around campus. When I met Nolen, I thought, "So this is the guy who makes those posters that I steal off the wall."

Nolen: Bruce's work struck me as smart and thoughtful. He had these big ideas, more so than other classmates.

The Double Dagger posters were Nolen's?

Nolen: Ideas come from both of us. Like the *Johns Hopkins 2007* poster over there—the whole thing was Bruce's idea. He drew the type, and I drew the rat.

Bruce: Nolen, who drew the explosions?

Nolen: I can't tell from here. I guess that's a good thing.

Does growing up in a small town lead to valuing community and human interaction?

Bruce: I think so. Nolen and I both come from a DIY, indie/punk background. Growing up in a small town means you don't have the same kind of access as people living in cultural centers. So when something does happen, everybody goes to it. When you see that people are passionate, that value is instilled in you.

Nolen: People have to work together with other like-minded people to create a culture that they want to exist in.

Bruce: A small town helps you empathize with other people. You can imagine yourself in somebody else's shoes.

How do you team-teach?

Bruce: When we do "good cop, bad cop," Nolen is the bad cop.

Nolen: Oh, I love being the bad cop!

Bruce: We usually teach at the same time because we are able to respond to students' work in a more rounded way.

Nolen: Something that I really like and think is good for the students, is when Bruce and I give contradicting feedback.

The Infantree

Ryan Smoker & Ryan Martin, Principals
Lancaster, PA

Creative Clash objective: run your own creative agency *all images courtesy of The Infantree*

home/work proximity of infantree design team

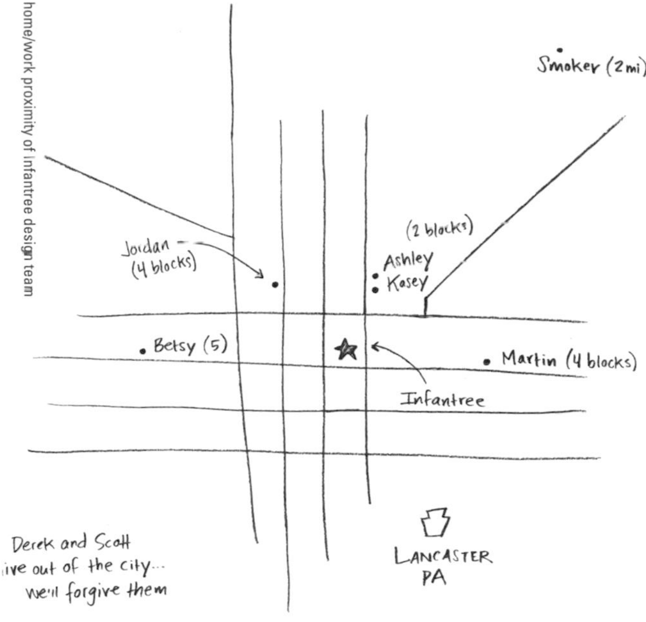

How do you form a footprint from Lancaster, PA?

In a small town like Lancaster, which is highly collaborative, your name is all you have. Make one bad move and word gets around really quick. We treat everyone with the respect and dignity that they deserve, no matter the size or budget of the client. A larger agency can fire someone for their failures or even move operations elsewhere entirely. Our employees' fault is our fault collectively. Our team succeeds or fails together.

Is there a link between group size + collaboration?

Yes, and I think we are scared of discovering the limit. If we manage more people and force collaboration, it's not really collaboration at all; it's a dictatorship. Eight of us fit into two cabs, and it's still a family. A 20-person weekend retreat to the lake house is too chaotic. When a team gets too big, sectioning starts showing up naturally. Most of our team lives downtown and walks to work. We couldn't develop an equal relationship with everyone if we had 20 people. We don't want to grow to such a size where we can't hang out anymore.

Do your clients come into the studio?

100% of our clients come into the studio. In fact, our process is that we force them to come see us. We bring clients into our office, pin up 300–400 images on a mood board, then allow them to react. This way they feel like they are a part of the creative process, because they are a part of the creative process.

How do you structure your team?

We always pick on agencies. Agencies will often go for the WOW, and once they win the account and clink their champagne glasses, the creative brief gets filtered down to the designers. And the designers are like, "What am I making? Why does this matter?" Our hope is to keep the levels as flat as possible. We want our clients to know our designers.

What do you look for when hiring designers?

The natural thing is to hire somebody that's just like you. But bringing in someone that works better, and different from you, is ideal. We thought of our first new hire as like a *Ryan 2.0.* We also have a personality test. I'd say that's even more important to us than raw talent. We've interviewed a lot of "rockstars." And that's great, but these kinds of people just want to have their names in lights, and I don't fault them for that. The Infantree doesn't have the space for a one-man band. We need everyone to wear hard hats and pitch-in to do the grunt work, even when it's not necessarily the most fun thing.

OK Go

Damian Kulash, Lead Singer
Los Angeles, CA

This Too Shall Pass Rube Goldberg machine, directed by James Frost

OK Go album cover *courtesy of Sagmeister & Walsh*

How similar are the processes of creating a song vs. a video?

It is a little different, but not as much as people think. One thing that is notable about our process is how much we invest in play, the theory of discovery, and trial and error. This process is more common in music than it is in filmmaking. The process for making movies is this: you sit at a desk and come up with the entire thing, basically. You get your script, your storyboard, what you shoot with, and by the time that you are at your shooting days, you have every minute planned. It is super efficient. That way, you aren't wasting resources by renting equipment that is just sitting there while you are still figuring out what you are going to do. The problem is, you are limited to the ideas that you have at a desk. There isn't much room for revision, or rethinking something, or heading off in a totally different direction. So our filmmaking process is to come up with the basic rules for the idea, to come up with the general area in which we are going to play, and then play in it for a very long time and only shoot after that. Music is sort of the same thing. Early in my career, I remember writing with a much more goal-oriented approach where I would think, for instance, why don't people write stadium rock anthems anymore? And then I'd try to write a stadium rock anthem. You can get good results that way, but it's hard to stumble across and discover things if you already know which way you are headed. So these days, my music process is a lot more like our filmmaking process, where we basically just play with the fundamental building blocks of songs: beats, chord progressions, melodies, sounds, etc., and look for those moments where something wildly emotional jumps out. Usually we are pretty surprised where we find that, because it's not at all where we would have headed. I find that a much more adventurous way to create.

How is collaborating internally on songs different from working with outside groups on a video?

It is a matter of scale. In either process, it's about getting people doing what they showed up to do, and having some people working on great ideas that they didn't see coming. When it's just the four members of the band, plus a producer, the boundaries between roles are pretty fluid. You need enough of a structure so that there's a place where the buck stops. But other than that, you don't need more structure. It's just people getting into the sequence of play. On a film set, you need slightly more structured roles, but we try to keep them as fluid as the projects can endure. For me, the key in collaboration is to be confident enough in what you believe, so that everyone in the room feels that the best idea is going to win. It's not about someone's ego or someone's preplanned anything. When really great things happen, it's usually because people lose their sense of authorship and chase whatever the best idea is. The band has been working together for 15 years, and we've known each other for much longer, so we don't need to convince one another of that anymore. We know that if someone chooses an idea that's their own, it's because they think it is the best idea, not because it's theirs. It seems like a small thing, but losing that sense of self is really important to good collaboration, so that you can be surprised by where an

End Love stop animation, directed by Lieberman + Gunther

The Writing's On the Wall *photo: Gus-Powell, courtesy of bbgun press*

idea goes, which means that you get the best of all of the brains, rather than just being together in a back room. When we are collaborating with another group, like when I am co-directing with other directors, there is generally a time period that we have to learn about each other, and get comfortable. The goal is to get to a point where people trust that the people making the decisions are doing so for the right reasons—not because they are being defensive. We've gotten pretty good at establishing a lighthearted work mode, where we all realize that we're here to make something awesome: "So let's just do it."

Do you prefer working in-person vs. online?

We get better results, and a lot more surprising results faster, when we are in proximity. But it is not always possible to do that. With music, we are not just there to jam things out. You need to have enough of a structure. I think of it as like cloud-seeding: you need the crystals there for the water droplets to hang onto. Usually, we work fairly separately for a few weeks or months, getting little snippets of things together that we can then throw in the sandbox and play with.

Are there specific considerations when working with volunteers?

From purely a logistical perspective, volunteers can be dangerous because if somebody doesn't show up one day, you're screwed. On the other hand, volunteers are working because they want to be involved in something that is really awesome. I love working with volunteers because they are there for the right reasons. The bigger the group of people working on something, the more likely people are going to have highly segmented roles, and the more likely people are to be jaded and professional about what they are doing, as opposed to naïve and full of commitment. It is always our role to try to keep everybody loving the actual thing that we are making. Our type of filmmaking is so uncommon that we sometimes get to see some old, jaded grip get really excited about something, because he's been shooting soap operas for the last 20 years. We encourage everyone to voice their ideas during all parts of the process. Because we do so many things in single shots, there is a pretty high chance that many times we won't actually get the thing that we came for. In normal commercial shooting, the difference between good and bad is a good and bad shot. If you are not very good at this, you are going to end up with something that is sort of B grade. If you have a bad day, you'll get three takes of each shot instead of six takes of each shot to edit from. But the entire thing depends on each shot at each moment. Whereas with our projects, it is all or nothing, so you get a much closer sense of team and investment. It is a very infectious type of creative lustiness: you are all going through this thing, and if anyone's part sucks, the entire thing falls apart.

Are your intentions for both mediums similar?

One thing with video is that the song already exists. There is an emotional roadmap that you're starting with that you don't have when you're starting a song. With a song, I feel like we are much more hunting in the dark, like we are in this room with the most interesting hallway out of here for a connection. In both cases, it is really about trying to give people a real emotional journey, and I don't think people usually think of video that way.

Animus Arts Collective

Preston Dane & David Ort, Co-Founders
New York, NY

Temple of Potential growable space to think, work, and interact *courtesy of Preston Dane and David Ort*

Your large-scale, public sculptures must demand collaboration.

Our core philosophy is about getting many minds together to create—thus all of our projects are inherently collaborative. That said, we distinguish it as not only collaborative, but more accurately, participatory. Artwork can be interactive, and a team can collaborate, but our work is not complete until a participant has collaborated with the artwork.

An example is the *Honey Trap*. A participant can mark the sculpture with writing or paint, and it was designed to be climbed on. The hexagon cells were designed to relate to the size of a human. So if we had displayed that piece on an island off in the distance where nobody could climb it, or in a way that didn't allow people to sit in the cells, or prevented people from having a conversation with their neighbor in the cell above them; while it would be beautiful to look at, it would be incomplete as a work of art. The most important collaboration is the one between us and our audience.

How do you account for so many perspectives?

We have some general stages for each project: Design, Fabrication, and Installation. After we decide upon a broad stroke for the overall design, we figure out logistics, like budget and design details. It really helps to work this out with other people, because if one person gets stuck on an issue, like how to attach something to something else, another person may have fresh eyes for the problem or a complementary skill set that can swoop in and save the day. Then we start making. There are a lot of repetitive tasks when working large-scale, so it helps that we all hang out and have fun while we work. The next stage is Installation. This stage is usually really hard, with long hours, and lots of heavy lifting.

It's important to think about who will be the participating audience early in the design process. People are leaving behind passive artwork and looking toward a new way of relating to the world through art. People seek to contribute. It is important that participants create a memory and leave their mark.

Does the Animus studio space impact the work?

Our studio can get hectic because of the size and materials that we use. Some projects are like building a ship in a bottle, then we take the ship apart and put it back together on site. This promotes collaboration in the design phase, because we have to be precise and calculated in what we choose to do.

Brothers Dressler

Lars & Jason Dressler, Co-Founders
Toronto, ON

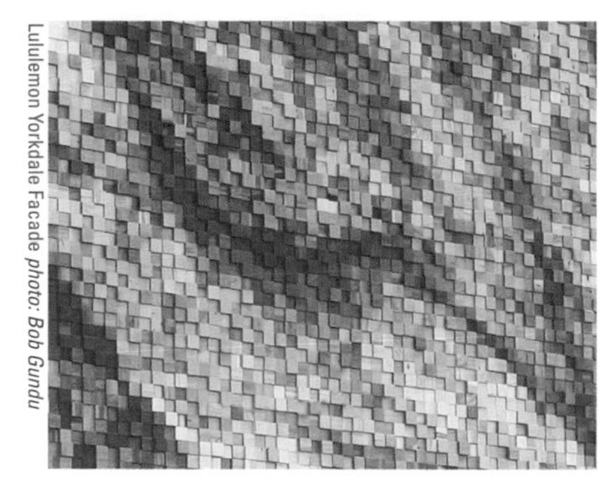
Lululemon Yorkdale Facade *photo: Bob Gundu*

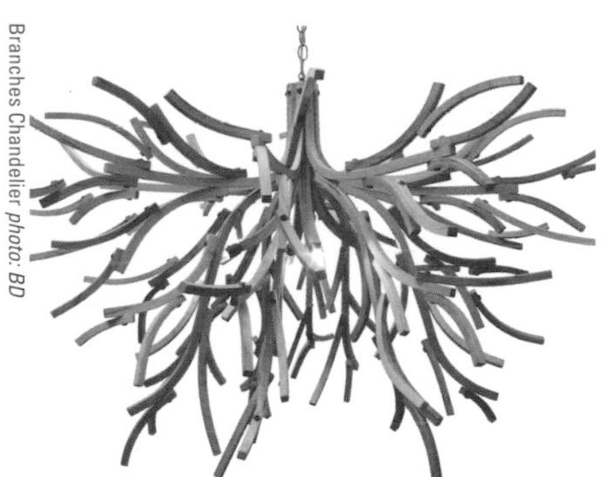
Branches Chandelier *photo: BD*

Lars (L) + Jason (R) in studio *photo: Dylan Macleod, all images courtesy of BD*

How do you go from brothers to co-workers?

Lars: We are twins, but we are very close twins. We started with Legos while growing up, built skateboard ramps, and made things for friends in our garage. The first collaborative piece happened when we were 18 and learning woodwork. Our family built a cottage in northern Ontario, and we made whatever we could using off-cuts. Jason studied mechanical engineering, and I studied chemical engineering. Jason went back to school for design, and we routinely discussed the dream of a joint business. After he graduated, we decided to make it happen. I quit my job and we started Brothers Dressler full-time in 2003.

How does your studio inspire collaboration?

Our work is integrated into our environment. We use a lot of found objects, so we're always collecting things, causing our studio to look a lot like a flea market; beaver-chewed sticks and leftover off-cuts are everywhere. Yesterday, someone trashed a bunch of smashed guitars, which is really sad. So we salvaged them and will see if we can do something with them, as they have beautiful wood. During the day, we have a number of people working with us and we spend a lot of the time with them to get things done. Sometimes Jason and I come back in the evenings for a second shift, after we've spent some time with our kids. That's when we do most of our creative work.

Do you have individual roles in the collaborative?

There is definitely a division of labor, but the creative work happens equally. Jason has a tremendous amount of experience in drawing and CAD, so the initial discussions happen together at Jason's computer because he's quick at making it look real in a short amount of time. Whereas, I often scribble ideas down and get materials in my hands right away. I also specialize in lighting and electrical. We go to different areas of our studio to work out details. Since we are very material-based designers, rather than just manipulating things on the computer, we like to get our hands dirty experimenting and enjoy the process of physically making things.

Why is craftsmanship important to a community?

The Maker Movement is thriving. Access to technology and information allows you to reach many people and let them know what you are doing. When something is locally crafted, and there are only 100 made, people are more inclined to want and value it. And by engaging with other companies that you respect, you make your business more worthwhile in the community.

Design-Student Groups

Jimmy Green, GD BFA Alum
University of South Florida, St. Petersburg, FL

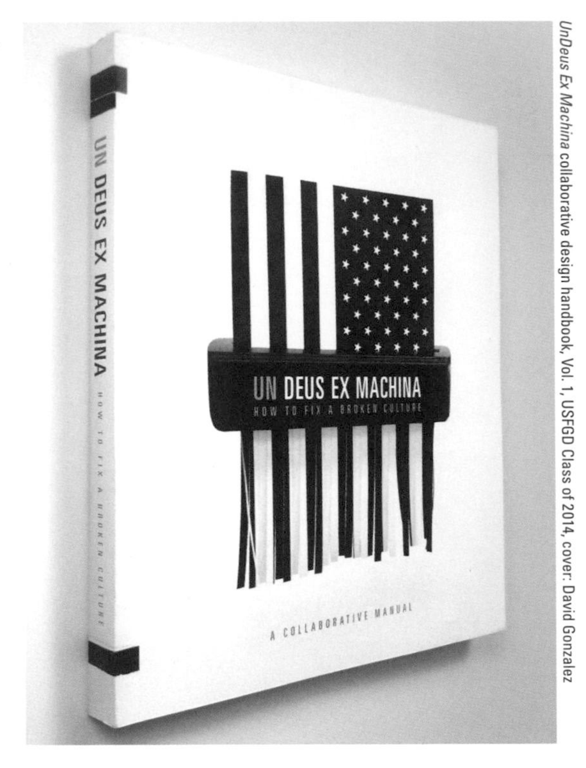

UnDeus Ex Machina collaborative design handbook, Vol. 1, USFGD Class of 2014, cover: David Gonzalez

How would you describe the collaborative group?

Our class was unlike any other that I've ever had. I don't mean "course." I mean, the group of people. From the beginning, everyone cooperated and was helpful to each other. There is an element of competition that is inherent to design students. It's similar to the arrogance seen in Ivy League students. Design students are in competition to be the best. That wasn't the case with our class, because the stronger students—which included me—agreed that they wanted to improve the program as a whole. We knew students were valuable teachers to each other. Those types of peer-to-peer experiences, things that you learn along the way, are often overlooked.

Does program size affect group dynamics?

Definitely. We had a small group. When you have less than 20 students and these are the same faces that you see everyday for a couple of years, people are more willing to invest in one another. Everyone attends the same classes at the same time. That changes people's mindset. People get to know one another because nobody's going anywhere. We were a family.

I've dabbled in a few other majors before settling on graphic design. And in every single one of those courses, people kept their heads down, did their own thing, then went on their own way. From semester to semester, you have new classes and have to start the process of learning new people from scratch. Unlike larger programs, our graphic design class was able to grow together because of the smaller structure. Even the competitive people, who thought they were better than everyone else, grew because of that dynamic. Students look out for their peers, respect the things their peers say, and are more familiar with the work that their peers are doing.

How did your class build a sense of cohesion?

Before we started in the graphic design program, a small group of four or five of us were friends and knew each other pretty well, especially Alex Maldonado and myself. Alex and I noticed that the class ahead of ours wasn't really doing much, nor were they improving, and we didn't want that to happen to our class. We took it upon ourselves to help steer the program in a better direction and imposed ourselves as a positive influence. We immediately made friends with everyone and invited the class to do fun things with us. I'd say meeting up outside of school was a big contributor to our bonding process.

For better or for worse, the administration at USF also largely influenced our "togetherness." We fought the University on some very serious issues. I don't think students stand up for anything as much as they used to in the past. As juniors, our schedule worked out really well, and alternatively, scheduling was what disbanded us as seniors. When we were juniors, our last class of the week ended on Thursdays at 6PM. Happy Hour usually runs from 4 to 7PM. All of us had drinks every single Thursday, and that allowed us to unwind as a unit. Also, every semester, we had an end-of-term dinner. We would rent out a room at a really nice restaurant, get dressed up, and have fun as a family. It's important for people to have time to enjoy themselves away from work, or in our case, the classroom. We knew everything about each other, which helped us trust one another. When we were seniors, our schedules destroyed that. Our last class let out on Thursday at noon. Everybody was like, "I want to go home, eat food, and take a nap." Some of them went off to work. We stopped being as friendly to each other. People quickly became annoyed by their peers, because we used to be close, so we felt at liberty to say certain things. After the dynamic changed, our group work suffered.

Ninth Letter

Maurice Meilleur & Brian Wiley, Art Directors & Faculty
University of Illinois, Urbana-Champaign, IL

Ninth Letter cover, Vol. 11, No. 1 courtesy of Ninth Letter

Ninth Letter is a joint venture.

Maurice: Yeah, it started ten years ago by the Creative Writing folks who came to the School of Art and Design.

What's the makeup of the group?

Maurice: There's an editorial end, and there's a design end. The design students work with submitted texts. The editor-in-chief is Jodee Stanley, who is in charge of the whole publication.

How many students are involved?

Maurice: There are typically eight to ten students in the graduate creative writing program who work as editors. On the design side, there have been as many as 25 students. Recently, it has been more like 13 or 14 students.

Why work with a smaller design group?

Brian: For students to really invest in it, they have to know that there is some sort of end product that they are going to be a part of. There is not that much room to meld 25 students together. It becomes more of a theoretical exercise for about 12 students, whereas if we can keep lower numbers, then there is a much higher chance of a practical application of their work.

Maurice: When there are more students, it is harder to get meaningful visual content into the issue.

How does *Ninth Letter* act as a course structure?

Brian: It's a special case. The graphic design students have to apply to be in the class. There are non-majors that we select, but they are by invitation only. It replicates professional practice a lot more than our other curricula.

Maurice: The course is an elective, so it's not meeting any specific requirements. It's a feather in the students' caps to be able to get in, and they are really happy if they make the cut.

What is the students' collaborative process?

Brian: It starts by identifying all of the different components of a journal. Then they break up into groups of two or three students per group, with each student in multiple groups. They propose visual treatments and design solutions for those identified components of the issue.

Maurice: They can work alone, but we encourage them to work together in small groups on the design.

Brian: They usually opt to provide joint solutions.

What is the timeline for the whole project?

Maurice: We put the issue to bed around Week 12.

Brian: This semester, we invited two design studios and one illustrator to actually come make a whole signature with the students: Studio-Set, Zut Alors!, and Adam Setala. So three of the signatures of the issue were a collaborative process between students and professionals.

Maurice: That was a much more concentrated timeline.

Brian: It was a gamble because of the time commitment. The intensity was really rough. The students are not used to working at that pace, especially with new people.

How does this compare to an internship?

Brian: The strength is that the students are allowed to fail. There is nothing at stake in some internships, so students don't take as many risks. In a safe environment, they take risks.

How does the interdisciplinary makeup of the group impact the process?

Brian: Graphic design students want to go from Point A to Point Z without anything between. They want to land on the finished product without trying multiple iterations and multiple prototypes and small-scale tests. One of our former industrial design students was used to prototyping and did mockups of the journal. She figured out a way to reveal a glyph character set on the fore-edge when you fan out the pages.

Odyssey Works

Abraham Burickson, Co-Founder; Ariel Abrahams & Ayden L.M. Grout, Members
Long Island City, NY

The Map is Not the Territory Carl's Joan of Arc Moment *photo: Ayden L. M. Grout, courtesy of OW*

What are the origins of the collaboration?

Abraham: My co-founder, Matthew Purdon, and I were discussing the problem of the ideal audience. We shared the common frustration that our work would be made for a particular effect. Some people would get it, but there was no way of knowing exactly how a person would receive the work. Celebrating that result began to feel like an excuse. We had this ideal reader/viewer/audience, but we didn't know who it was.

What goes into an *audience of one* performance?

Abraham: We have an extensive application process to help us choose and get to know our participant. This involves filling out a long questionnaire. We call references and conduct live, videotaped interviews with our finalists. This work is split between the members of the Structure Team, which is a three to six member group that designs the overall vision of the piece. Our group work is about seeing a person. Everyone draws a picture and presents it to the group. It's our way of saying, "Look, this is what I see. Can you see it too?"

How do various skill sets complement each other?

Ayden: We cover a lot of disciplinary ground in our core team, and we also work with psychoanalysts, composers, dancers, designers, storytellers, and computer geeks. Once all of this basic research is done, the structure team retreats somewhere outside the city between four and seven days to eat, breathe, and sleep in the participant's world. In making such large-scale, ephemeral work, it has become a really crucial part of the process to create a diagram that delineates all the strata of a ±48 hour performance. Abraham thinks about the arc of the experience for our participant, and as an architect, he really likes to chart the many layers of each Odyssey experience.

Ariel: We are not afraid to mix dance, theater, visual art, and music together. Even with a large network of potential collaborators, the crew is tight, and each artist's unique talents come into play. A magician might volunteer, but it's not about doing a magic trick; it's about the relationship to our participant.

Do you see collaboration in the arts shifting?

Ayden: There is more demand for collaboration in contemporary art. I see fewer artists sticking to one craft or a single medium. Collaboration has been a gateway to interdisciplinarity. Not only that, but it has been an invitation to expand the scope of our vision. It also advances my individual practice; if I donate time to collaboration, I need to allot time to myself in my own studio.

Temporary Services

Brett Bloom & Marc Fischer, Co-Founders
Chicago, IL & Copenhagen, DK

page from *Prisoners' Inventions: Three Dialogues* collaboration with Angelo *all images courtesy of TS*

Booklet Cloud installation of TS publications

Group Work perspectives from artist practitioners

distributing publications via free newspaper dispensers

What led to working together?

Marc: Temporary Services started in 1998, in Chicago. Brett and I met in grad school, and we met the others through the work that we were doing. We were looking for ways to expand the audience of experimental culture by doing event-based things. Publishing was incorporated from the beginning when we made free booklets for our first exhibition.

Brett: During grad school, I didn't really care for the things that Marc was painting, and I'm sure he didn't give a shit about what I was painting. Two years later, I was making publications that could be distributed through a network of free newspaper dispensers that I had set up. We were interested in getting out of exhibition spaces and into public spaces where we could get direct feedback. Friendly audiences go to art openings and tell you that "you're great," "congratulations," and "go have a glass of wine." We wanted more critical engagement.

Do you still meet-up to run projects?

Marc: We see each other at least a few times a year. Our publishing has always consisted of files being passed back and forth. Fortunately, as we have grown further apart geographically, technology has expanded.

Brett: But we still have to do group maintenance. I don't know what Marc is going through unless he shares that and vice versa. We have to take the time, once in a while, to overtly check in and see where each other is at. It's a chore that you wouldn't have to deal with when you're living near one another.

What have you learned by working together?

Marc: It is hugely important to create social structures that take care of people. Everything in school is set up against collaboration—even collaborative classes still feed into a system that produces individual degrees or individual consumers of education. Finding ways to deal with conflict and mitigating it, and finding healthy ways of reflecting everybody's needs within a group by building these kinds of structures, is hugely important. We didn't start out that way initially, and we ended up alienating people that we really cared about. It is a stupid thing to have to go through, but guidelines help.

Brett: Telling students that any assignment throughout the entire semester can be done collaboratively, is helpful. They can work with people outside of the class and bring them into the discussion of the work. A lot of people are not encouraged to even try that in their education.

Ben Kiel+Ken Barber

Ben Kiel: Typefounding
Ken Barber: House Industries

Cortado Script *courtesy of Ben Kiel*

Is type design primarily collaborative?

Ken: Looking at the history of metal type, persons credited with creating typefaces were rarely the only ones who contributed to their manufacture. Usually the designer would conceive of the aesthetic blueprint, while a punchcutter carved the steel "punches" used to make molds from which the type was cast. It was uncommon for the designer to be personally involved in the cutting and casting; others were responsible for physically producing the actual type. It's a rather new phenomenon for the process of typeface design to be a solo effort. Even today, I prefer working with others because the collaborative process allows greater potential to find better solutions.

Ben: Yeah, I think there has been a shift, but actually that shift is reversing. There is a time and a place in the typeface design industry where you can still do that; you can design a typeface sitting in your room and not talk to anyone. Once software became affordable, the means of production and distribution became accessible to all. But at least in the type industry, I think the trend is different now. It's one thing to do a poster font that requires only a particular set of characters by yourself. However, if you are doing a superfamily, there are all sorts of things to cover. We've essentially bridged the technical point where the needs of type design outstretch the ability for one person to do everything themself. I'm thinking of an example like Tal Leming and his recent Balto typeface. He drew the entire thing. He kerned everything, but he didn't do the hinting. And I think that's going to be the case more often. Large character sets aren't just English, web fonts need hinting, and superfamilies are huge tasks.

Designers doing everything themselves was a business model that was set up upon hope: a designer designs something and hopes it has success. They keep overhead low by doing as much work themselves, in order to retain as much profit as they can. So designers had to learn all of these skills, but people are realizing that they can't do it all, despite being told they had to.

How does the House studio facilitate collaboration?

Ken: House Industries co-founder Rich Roat likes to describe the studio as having an important "rollover factor." Our office is somewhat small, which allows us to roll over to each other's desks and offer instant feedback. And despite working remotely a couple days a week, we're always wired, messaging and exchanging files. Most importantly, we leave personal agendas at the door, focusing on the ultimate goal rather than letting egos get in the way.

Ben: There is a very strong culture and collaborative effort at House. Everyone is used to being able to get instant feedback on their work—that's just not the same over email, time-zones, and iChat. A good example of remote collaboration is the work that I did with Jesse Ragan on Carlstedt Script, which later became Cortado Script. Cecilia Carlstedt drew out a bunch of characters as a starting point. Then, Jesse and I spent a lot of time messaging, exchanging emails, and talking on the phone to transform the lettering samples into a functional typeface. We split up the character set: Jesse did the lowercase, and I did the capitals. Then we swapped. We exchanged files through Dropbox. For typeface design, you can't just divvy up parts of a character set indefinitely. But it makes sense for designers to take a segment of the project at a time, then hand it over to the other person to rework and make edits.

How did you team-teach typography?

Ben: Ken teaches to his strengths. He enjoys tightly structured assignments. I'm a little looser than that; I'm not as worried.

Velocipede Salon logo *courtesy of ©House Industries*

Typographreaks: Masters of Typography artwork for *Fast Company* article on House Industries *courtesy of ©House Industries*

TYPE SUPPLY

Tal Leming, Founder & Former House Industries Type Designer
Baltimore, MD

Balto revisions stacked *courtesy of Type Supply*

Catalog of Neutraface cover *courtesy of ©House Industries + Type Supply*

Ohm Bold + Torque Book + Marigny Book *courtesy of Type Supply*

What is the design process for a typeface?

I either have a client come to me with a particular project that requires a typeface, or more commonly, I create typefaces for myself to sell via my website to designers. On that stuff, I am often by myself, which is a hindrance.

Why do you say that it is a hindrance?

I used to be a graphic designer, and that process has a faster feedback system. Typefaces are more abstract. It is like making a new kind of hammer: Will anybody use it? Without someone to bounce things off of, I second-guess myself all of the time. I don't have a lot of self-doubt, because that sounds really depressing, but I spin a lot. As a perfectionist, the answers to ambiguous questions are hard to find, because there is no perfect answer. I swear I am not crazy.

How did you get into web standards?

Kerning is a nightmare, so I wrote a kerning editor, and I needed a file format to use with it. Meanwhile, some friends, Erik van Blokland and Just van Rossum, were looking for a way to collaborate on drawing type, so Just invented the Glyph Interchange Format. When OSX updated, lots of in-process fonts quit working. We jokingly said that there should be a file format that isn't dependent on a particular application.

We wrote this file format called The Unified Font Object, or UFO, which Erik came up with because he's funny. Erik wrote a tool for interpolation, meaning you can draw the lightweight of the font and heavyweight and calculate the ones in between with math. Frederik Berlaen, a type design student at the Royal Academy of Art in The Hague, wrote a UFO font editor that we all use called RoboFont. There is this whole ecosystem of designers writing scripts and putting them on GitHub and extending RoboFont.

Then, Erik and I jokingly, and completely stupidly, decided that we would negotiate a way to use type in browsers. We did that by writing a file format, because we had the experience of doing the UFO. We combined it with a proposal from a Mozilla computer scientist, and after lots of back-and-forth, it is now officially a web font standard. But it all came out of trading stuff.

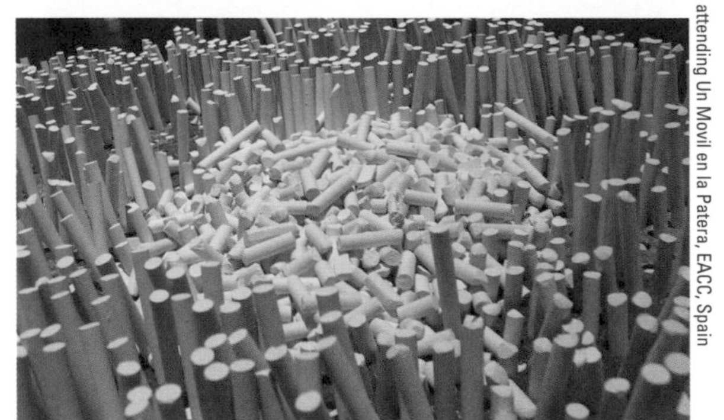

Chalkboard installation provokes interaction from visitors attending Un Movil en la Patera, EACC, Spain

Pixel Wall kinetic picture board, Queen Elizabeth Olympic Park *all images courtesy Tomato*

What makes the Tomato process unique?

Tomato has been together for 23 years and still has a lot of the original people. One of the key things for us is we completely dropped hierarchy. We have quite a wide age range. It's unique, because you don't know how the situation will evolve due to the lack of hierarchy and different ages and different experiences. It's a very fertile ground because of that.

Are concepts individually or collaboratively created?

It really depends on the project. We did a big sculpture recently for the Olympic Park in London. The two people that worked on that thought about it independently, pulled their ideas together, colluded on that, went away, developed some more, came back together, and then spent the next 12 months working closely on everything. I think that's a natural flow. You can't be sitting in each other's pockets all the time. People have to feel a certain amount of freedom to be able to do what they do.

Ideas do come in collaboration, definitely, but I think you have to have a fairly fertile, individual mind to make those things happen. That's a crucial point to make in a successful collaborative team: the group is the main focus, but individuals are allowed to have their own identity at the same time.

Do you bring clients into the studio?

People come in. We recently had a bunch of Japanese clients come all the way over from Japan; they came to our studio and sat down on a bunch of fatigued seats and stools. They had a cup of tea while our dog was running around interrupting them and wanting to play with them. I think it is important that clients see where and how designers work in the raw, and to know what they are going to get, to some extent. When clients go away, they can imagine you and what you are doing for them. It's beneficial for them to see us working in our space because they don't live and work in isolation themselves, so they are often open to what other people do.

What do clients see in your studio?

It is like a home. I think people get the idea that it is very communicative and quite a relaxed world. That helps people make things that come to the fore. I've worked in very formal offices with notebooks around, but no energy; it's devoid of any positive cultural influence. I see how studios impact work ethic. Sometimes that occurs in ad agencies. They are designed to be like that. They are more like machines for making a certain type of product. I guess they do that very well.

TelephoneMe MK12 original film *courtesy of MK12*

What makes the MK12 process different?

Our studio is very much an artist collective. We all collaborate with each other on most every project, internal or client-based, from start to finish. The four of us who started the studio 13 years ago either graduated, or dropped out of the Kansas City Art Institute. At school, we'd help each other out on our own respective projects, lending a hand for the grunt work when needed. We also teamed on collaborative works, usually of the kinetic, animated, experimental nature. We'd help build upon initial ideas that we were all coming up with and expand them into workable projects. Each of us had certain strengths, like 3D modeling or compositing. We've always played to our strengths, but we share the role of Creative Director.

Everyone can participate on all levels of a project, from initial ideas to storyboards to rotoscoping to compositing and any other animation-needs after that.

We have no idea if we're doing it right. We're based in Kansas City, so our competition isn't exactly around the corner to check out other facilities, clients, or tricks. We just do what we do. We like what we make, so that's why we continue.

How does your studio inspire collaboration?

The biggest nod toward collaboration within the studio is that we have no walls between our workstations. All in all, the space is around 6,000 ft^2, counting the front space that serves as a conference room. Rather than a long table to fit the eight of us, we refurbished a smallish airplane wing. We used to have a ping-pong table, but it was destroyed through lots of playing.

What is the arc of an MK12 project?

Each of us have our strengths, but as a group we've become pretty balanced in terms of jumping in and around our production line. We're able to pick up where others leave off. It's pretty organic. Amongst the eight of us, our shared skill sets can accomplish almost everything in-house. A lot of the time, we're the background talent in our internal work. We're cheap, and we're usually available.

Does authorship matter in these group projects?

Everyone feels a shared sense of authorship with the work. Work is always credited to MK12: directed by MK12, written by MK12, animation by MK12, VFX by MK12. Since we're all creators of the work, it's easier to have it all fall under the studio moniker. There's also something nice about the anonymity.

!NDIVIDUALS

Luke O'Sullivan, Colin Driesch, & Winston MacDonald, Co-Founders
Boston, MA

happy family portrait *courtesy of Ind!v!duals*

construction process *courtesy of Ind!v!duals*

Can you talk through the origins of !ND!V!DUALS?

Winston: We played in the woods and made treehouses when we were kids. We knew each other for a long time, and that's a big part of it; to be able to coalesce a collaboration like this means everybody has to be involved in everybody's space.

Colin: Growing up in the woods. Being sarcastic. Humor. Bouncing a lot of stuff off of each other. Our main influences are each other. We riff off of each other and do our own thing.

Is there an !ND!V!DUALS process?

Luke: Every single thing we do starts with drawing.

Winston: It starts with listing ideas first. We sit around the kitchen table throwing out ideas left and right. Then we move to sketching with Sharpies; Luke and Dom are pretty incredible.

Colin: We have to show clients past projects with the Sharpie drawings just so we can reassure people that they can trust our Sharpie drawings, and that it will come out okay.

Luke: We should send our Sharpie drawings to China and get high quality illustrations of our work mailed back.

Colin: Yeah, there are a lot of trust issues with galleries and curators anymore, so we have to promise them that it always works out as we draw it up. People get nervous. Nobody in Boston is doing the weird wacky stuff that we are doing.

Luke: There is a loose, organic way of creating these projects. People want to know what the processes are and how we put all of this stuff together. Honestly, it's a couple of guys sitting around chatting and sketching stuff on the back of pizza boxes, napkins, or restaurant placemats.

Winston: From there, we take that lid of the pizza box with its pretty little sketch, and we break off into groups. We have a list of things that we need to build for the concept. People pick what they want to execute. It goes straight from Sharpie drawings to buddy groups for the construction.

I talk a lot about it with video work. It's more than just having the extra hands in a collaboration, it's a springboard for having those extra opinions. Having all of those motivators around you means that you are receiving those ideas constantly. A lot of artists will contract the work out when they do a large-scale project. It's not hard to find interns and labor, but having the five of us for physical labor is exciting. The important part of it is getting five people working conceptually together, which makes things happen very quickly. It gives you confidence and moral support to bounce ideas back and forth.

MICA Center for Design Practice

Mike Weikert, Director & Ryan Clifford, Assoc. Director
Baltimore, MD

Discover Greensboro Traveling Spring Break, AL

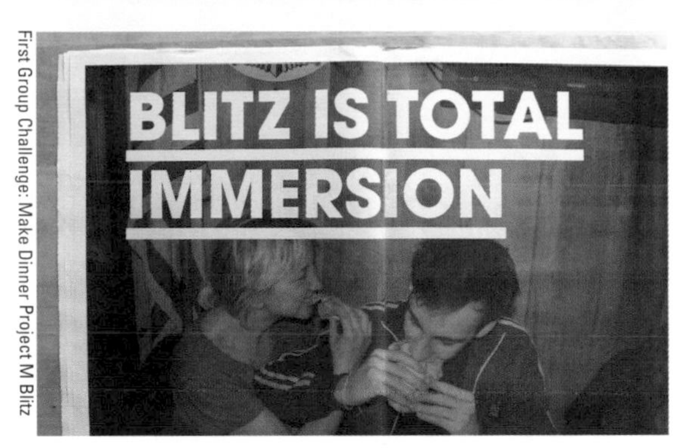

First Group Challenge: Make Dinner Project M Blitz

What are your roles as Social Design faculty?

Mike: That touches on the social skills that design thinkers need to bring into collaborative environments: knowing when to lead and when to be led. Collaborators have to be nimble in order to play different roles at different times. New students often think that means they have to be the expert, or they have to succumb to whatever the community wants. But in reality, it is the ability to absorb different processes and adapt to different contexts.

Ryan: Mike and I are not together in a classroom in the same capacity. Mike is a resource. He comes in at different points during the semester to challenge the students, facilitate the project with a fresh viewpoint, and to make sure everyone is on task. Whereas I am with the students full-time. With the Project M Blitzes, there is a huge element of trust that the students seek from me. I collaborate with John Bielenberg, and we've been working together for quite awhile. Mike and I overlap in our philosophy, but we have our own approaches to making that work.

I want students to look at the value of their skill sets and see how they can best contribute. We don't define "meaningful." We expose them to a ton of different approaches. A big aspect to learning about collaboration is having students see other designers come together with differing approaches.

Why do you two work well together?

Ryan: We have similar backgrounds. We were both working professionally doing branding, and we found that it was not as rewarding as it could be. We have complementary skill sets. Mike is good at seeing the big picture and developing strategies. For example, the majority of the work that we do is based on relationships that Mike has built. He created the CDP, now called Practice-Based Studio. He directs the MA in Social Design. What I bring to the table is my teaching ability. I love the classroom experience; I like using what I learned about design to address social issues with students. I like group dynamics.

How is collaboration integral to social design?

Mike: We see it through two lenses. The first is that everything we do is driven by a social issue or problem. The second part of it is that we believe that social design is not a solitary graphic designer working on a social issue. It is a collaborative, interdisciplinary approach. A social designer is not a social designer unless they are social. Collaboration is embedded in the core of social design.

What is the process for a CDP project?

Ryan: The CDP is process-oriented. There is a core design challenge, but we don't have an agreed-upon outcome. Our partners and audience are heavily involved in the process.

Mike: Projects often span multiple semesters. The initial 16-week engagement with a partner gets us to having ideas and options.

Ryan: At the end of the semester, students share pitches. The client-partners usually want to continue the relationship.

Does emphasizing process eliminate anxiety?

Mike: If there is a deliverable that is due at the end, people go back into their little silos of expertise and comfort and complete the task on their own. We have partners who are writing us new grants for year-long projects because they have bought into the value of the collaboration and the collaborative process. They no longer see design as just an outcome.

When you are able to fondly look back upon an endeavor, challenging moments are suppressed by the impact of good work. There is never a collaborative experience that doesn't suck at some point along the way.

Nancy Skolos+Tom Wedell

Interdisciplinary Design Faculty, RISD
Canton, MA

event + competition posters construction process highlights photography + design *courtesy of Skolos-Wedell*

How do you handle collaboration in an art school?

Nancy: RISD is still based on a deep disciplinary structure with 19 different areas. We catalyze collaboration through a menu of courses. We recently came up with a concentration called Nature–Culture–Sustainability Studies where students are given a list of courses that relate through architecture, philosophy, etc., and this helps give the students a thematic path. Also, we just completed a house for the Solar Decathlon, which involves both design and fine arts disciplines. We called it the *Techstyle Haus.* Textiles made fabrics for the interior, Glass made glassware, Ceramics made the dishes, Furniture made the cabinetry and furniture, and there was art in the house from all of the students.

Tom: One of the advantages at RISD is that Brown University is right there. I taught a course mixing Writing students at Brown with Industrial Design (ID) students to re-examine the purpose of the book and find a variety of possible forms of the book.

Nancy: It was called Built Thought. The course structure itself was a collaboration, and there were three teachers: one from Brown, one from ID, and Tom from Graphic Design.

What is the collaborative process in your own work?

Nancy: Most of our work is an honest 50-50 split, but some of it can be done 90% by one person and 10% the other. I think that the secret to our successes is that the goal of the project takes the first priority, and we develop our work towards that. We are not working for ourselves; we are working for the result. A lot of times, it starts with sketching. We talk about stuff, and then we sketch stuff. It's funny, because we go to this Japanese restaurant, and we always draw on the back of the placemats. Tom can draw upside down, because he is so used to working in a view camera.

Tom: It is one of our favorite restaurants, so they automatically bring us placemats, and we draw immediately, coming up with ideas and writing words in the margins. They must think we're crazy, but they're used to it now. This is how collaboration works.

Do you work with others?

Tom: If you want to see the challenges of collaboration, build a house. Having designed a number of interior products, Nancy and I have worked with couples that can't decide on anything. It's always a problem of, "I'm right and you're wrong." Our job, as the "psychologists," is to look at the goal and see what really makes a successful project. If we can do that, then it begins to work. If we can't do that, then it becomes contentious.

ADColab (RISD Grad Design Course)

Brooks Hagan, Co-Founder & Faculty
Providence, RI

Sensory Deprivation Suit Mimi Cabell + Jamie Foster, project: Failure is the goal *courtesy of Brooks Hagan*

What are the origins of Interdisciplinary Collaboration in Theory & Practice (ADColab)?

It was an experiment that emerged from co-founder Julian Kreimer and I attending grad school together at RISD. We interpreted some of Julian's drawings with Jacquard Woven Fabrics, which is my medium, for a painting class. A couple years later, we decided to formalize that experience and multi-perspective process. Julian is a painter with an interest in design. I am a designer, with a fine art lens. We got a grant to run it for one year. It is very resource intensive. For RISD, it's like a double class of 22 or 24 students from different areas and two faculty members. In the first year, we did four trips to New York, had visiting critics, an ambitious reading list, and four collaborative studio projects. We were able to tone the quantity down the second year, and make it pertain to topics more relevant to both art and design: permanence, mechanical reproduction, and art vs. design. We asked students to work in groups to prepare a response and to guide the discussions of the readings. Now, it has transitioned into a new iteration. Originally, it was supposed to be an artist and a designer co-teaching with students from many different disciplines, but I also taught it solo for two years. It was a fun class to teach, because I met people from different departments with different working processes, and I witnessed incredible results of being in a shared space. The hardest thing was anticipating the organic flow of discussion without derailing it. Two faculty members in the room can overpower the students.

Do you teach collaborative theory?

Projects become team-based. The finished results aren't quite as important as the process or experience the students have. They learn to work with others and form deep bonds.

But collaboration is never addressed head-on as a thing, except for the beginning of the class, when we do some intro readings for the designers to think about the art way of seeing, and for the artists to see the design process. By the second week, we are down in New York visiting artists and designers, studios, workshops, shows, and installations. The students go on these trips together three times during the semester, in addition to doing the collaborative projects. We throw them into the collaborative experience and assume that they will work it out. Sometimes it doesn't work out. There are people who don't get along, but that is informative as well.

Do all disciplines naturally work well together?

Yes, but it can be a challenge if somebody is too technical about their process, so we get them to look at the *play* inherent in what they do. People discover that they practice one discipline, but have talent in another area as well. The first assignment we give is called *Tech Thing*. Students make an object that satisfies the formal standards of the disciplines that they represent.

How do you solve for groups that don't work?

The Failure is the Goal project is supposed to truly be a failure. The group may not get along well and hand in unresolved work. That becomes the subject of the discussion. I help push their projects, but for the most part, they just have to deal with it.

Does emphasizing failure help experimentation?

Yes, I hope so. People often finish that project and are very confused and not happy with what they have done. But that's part of the process, and they usually approach the next thing with renewed intensity. The fine artists are always very comfortable with the failure thing, like "Yeah, we can do it." But sometimes, the designers just can't accept failure. It's too challenging of a process for them.

Yellow Bird Project

Matt Stotland & Casey Cohen, Co-Founders
Montreal, QC

Yellow Bird Project logo design: Dynamo *all images courtesy of YBP*

Dresden Doll Amanda Palmer's contribution to the YBP collection donations to the Syrian Refugee Crisis

What is the Indie philanthropy/collaboration link?

Matt: Collaboration is what made YBP possible. We've worked with designers who could be million dollar enterprises if they had the network, but because of their small scale, they were willing to work with us. Combining resources expands options.

How does YBP benefit artists/musicians/charities?

Matt: They are all dependent on each other. Our collaboration with Andy J. Miller on *The Indie Rock Coloring Book* and *The Indie Rock Poster Book* are great examples of everyone's skills layering on top of one another. Andy had all these amazing ideas, we had the fans and reach and marketing skills, and Chronicle Books had the distribution. We were able to take this passion and turn it into an impactful commercial success.

How do you collaborate with musicians?

Casey: The selection process is fairly arbitrary—we go after our favorite bands. Not everyone gets to choose who they work with, but we do. That's one of the perks. Once we've chosen a band, our first job is to marry them with a designer or illustrator whose style fits with the band's overall aesthetic; that is, if there isn't a band member who can design the t-shirt by themselves. For example, the *Dry the River* t-shirt was designed collaboratively between lead singer/guitarist Peter Liddle, and Jonathan Lindley, our Art Director at the time. In an email, Liddle said, "I really like the idea of using old books to say something about the weight of history and how we all labour underneath that." So they organized a photoshoot at Turton Tower, which is an historic building halfway between Bolton and Darwen. The design includes a photo by Lindley that shows the silhouette of a boy facing forward, balancing history books on his head. The image is based on their song "History Book" and depicts a Victorian technique used in schools to correct children's posture. The boy sitting on the chair is actually the band's guitar player, Matt Taylor.

How did you get started?

Matt: We met in high school, in Montreal. After graduation, we spent a summer together in London where we got into all sorts of music and started going to concerts together. The Montreal music scene was really blowing up and getting international recognition at the time. We thought of combining all of our concert and band t-shirt passions into one, and YBP is what came of that: giving back, working with our favorite musicians, and producing a product that we would buy.

The Copycat

Rob Brulinski, Co-Author & Former Resident
Baltimore, MD

The Copycat warehouse converted to residential compound courtesy of Rob Brulinski + Alex Wein

The Copycat photo essay on artist commune courtesy of Rob Brulinski + Alex Wein

How do you characterize The Copycat residence?

The people who visit are generally art students, or people who are there to see something, like a show, or a gallery space, or a theater performance. The people who live there actually vary greatly. There are some students, but usually it's older people. Like, there was a retired social studies teacher and his whole room was full of instruments. This dude is retired, in The Copycat, smoking joints, and playing all the music that he wants with a plethora of instruments. It's beautiful. Another older guy makes jewelry and has six employees in his space. While he was making jewelry, they were building a design school and tearing a hole in his wall! There was a doctor that lived there just because. The weirdest, was the guy who sold real estate.

Why do you keep returning to it?

It's the sense of duty and the sense of freedom that you get. As an artist, I can go there and meet people who are taking their life just as seriously as I am, or just as not-seriously as I am. You never know what you're going to wind up with. When I lived in The Copycat, there was never a dull moment. My neighbors were magicians. And I never saw my other neighbor, but I knew she sold sex toys out of her apartment, online. That was like, her thing. I thought it was hilarious. I'm sitting next to a dildo factory and a magician, and that's such a small part. I like going there because it reminds me to be creative, to research, to continue being an artist, and to continue being myself. You don't need to replicate what is on Instagram or what someone in New York is doing, because you are not there. Wham City moved to Baltimore to live in The Copycat because it was cheap as fuck. And then they took that idea further, and said, "We are going to be artists and share ideas without any objections or any prejudice." It would be weird if you wanted to be an artist but didn't do something there, because it's been around so long.

How does the environment impact the work?

You are allotted your space. If you are living in a huge space, and you have eight roommates, and two of those roommates bring four other people in illegally, now you have a lot of mouths to feed and a lot of rent to pay. And if you don't have a real job, you better find some fucking shows. And you're going to start talking to some other people about doing stuff in that building. There is a lot of connectivity going on in a space like that. It forces you to open your doors to strangers for the sake of art, a couple dollars, and a good time.

Bibliography

AIGA Staff, and Nicola Bednarek (ed.), *Fresh Dialogue 6: Friendly Fire (New Voices in Graphic Design)*. New York: Princeton Architectural Press, 2006.

Anderson, Chris, and Michael Wolff. "The Web Is Dead. Long Live the Internet." *Wired,* August 2010.

Aoki, Keith, James Boyle, and Jennifer Jenkins. *"Bound By Law? Tales from the Public Domain."* Creative Commons. Durham: Duke University Press, 2006.

Asimov, Isaac. *I, Robot.* New York: Doubleday (Random House), 1963.

Barringer, David. *Emigre No. 68: American Mutt Barks in the Yard.* New York: Princeton Architectural Press, 2005.

Barthes, Roland. *Image–Music–Text.* Translated by Stephen Heath. New York: Hill and Wang, 1977.

Barry, Max. *Company.* New York: Doubleday (Random House), 2006.

Bauer, Erik. "Method Writing: Interview with Quentin Tarantino." *Creative Screenwriting,* August 2013.

Bertalanffy, Ludwig Von. *Robot, Men and Minds.* New York: George Braziller Inc., 1967.

Bichlbaum, Andy, and Mike Bonanno. *The Yes Men Fix the World.* New York: HBO, 2009.

Bierut, Michael. *Seventy-nine Short Essays on Design.* New York: Princeton Architectural Press, 2007.

Book Sprint Team. *Collaborative Futures.* Re:Group, Eyebeam. New York: Lowercase Press, 2010.

Borden, Louise. *The Journey That Saved Curious George: The True Wartime Escape of Margret and H. A. Rey.* New York: Houghton Mifflin, 2005.

Bosman, Julie. "Selling Books by Their Gilded Covers." *The New York Times,* December 2011.

Bosman, Julie. "Barnes & Noble Won't Sell Books From Amazon Publishing." *The New York Times,* February 2012.

Boyd, Andrew, and Dave Oswald Mitchell. *Beautiful Trouble: A Toolbox for Revolution.* New York: OR Books, 2012.

Briggs, Katherine, and Isabel Briggs Myers. *Myers-Briggs Type Indicator (MBTI) Instrument.* Center for Applications of Psychological Type (CAPT), 1975.

Bruni, Frank. "The Wilds of Education." *The New York Times,* September, 2014.

Buckley, Christian. "Making A Case For Social Collaboration Tools." *Wired,* March 2013.

Burroughs, Miggs. "Interview with Graphic Designer, Paul Rand" Connecticut: Miggs B. on TV, 1991.

Burton, Summer Anne. "40 Inspiring Workspaces of the Famously Creative." *BuzzFeed,* April 2013.

Cain, Susan. "The Rise of the New Groupthink." *The New York Times,* January 2012.

Carlin, George. *Napalm & Silly Putty.* New York: Hyperion Books, 2001.

Chin, Andrea. "Stefan Sagmeister Interview." *designboom,* May 2006.

Colbert, Stephen, Jon Stewart, Ben Karlin, and Tom Purcell. *The Colbert Report.* New York: Comedy Central, 2005–2014.

Conderman, Greg, and Bonnie McCarty. "Shared Insights from University Co-Teaching." *Academic Exchange Quarterly,* Winter 2003.

Davis, James. *Interdisciplinary Courses and Team Teaching: New Arrangements for Learning.* Westport: American Council on Education and The Oryx Press, 1995.

Davis, Joshua. "How a Radical New Teaching Method Could Unleash a Generation of Geniuses." *Wired,* October 2013.

Debord, Guy. *Society of the Spectacle.* Paris: Editions Buchet-Chastel, 1967.

Dick, Philip K., *Do Androids Dream of Electric Sheep?.* New York: Del Rey Books (Random House), 1968.

Didion, Joan. *The Year of Magical Thinking.* New York: Alfred A. Knopf (Random House), 2005.

Dobbs, David. "Glowing Maps of Scientific Collaboration." *Wired,* January 2012.

Doctorow, Cory. *Down and Out in the Magic Kingdom.* New York: Tom Doherty Assoc., 2003.

Doctorow, Cory. *Little Brother.* New York: Tom Doherty Assoc., 2008.

Doctorow, Cory. *Makers.* New York: Tom Doherty Assoc., 2009.

Doorley, Scott, David Kelley, Scott Witthoft, and Hasso Plattner Institute of Design at Stanford University. *Make Space: How to Set the Stage for Creative Collaboration.* Hoboken: John Wiley & Sons Inc., 2012.

Dover, Caitlin. "Design Couples." *Print Magazine,* June 2010.

Dragilev, Dmitry. "XBox One, Netflix, Charles Schwab: Why Consumer Collaboration is Key in Business." *Wired,* June 2013.

Duru, Ruxandra. "Type Foundries Today: A Study of the Independent Type Foundry Industry." Self Published, 2011.

Ellis, Warren. *Transmetropolitan, Vol.1–10.* New York: Vertigo (DC Comics), 2009–2011.

Forsyth, Donelson. *Our Social World.* Pacific Grove: Brooks/Cole, 1995.

Frana, Maura, Leigh Mignogna, and Liz Seibert. *Five Conversations on Graphic Design and Creative Writing.* New York: Pratt Institute, 2013.

Friedman, Thomas. "Average is Over." *The New York Times,* January 2012.

Gaiman, Neil, and Dave McKean. *Signal to Noise.* Milwaukie: Dark Horse Books, 1989.

Gaylor, Brett. *RIP: A Remix Manifesto.* Victoria BC: Open Source Cinema, 2008.

Gertner, Jon. "True Innovation." *The New York Times,* February 2012.

Gertner, Jon. *The Idea Factory: Bell Labs and the Great Age of American Innovation.* New York: Penguin Group, 2012.

Giampietro, Rob, and Ian Albinson. "School Days." *Graphic Design: Now In Production.* Minneapolis: Walker Art Center, 2011.

Gonchar, Michael. "Do You Perform Better When You're Competing or When You're Collaborating?" *The New York Times,* October 2012.

Goodman, Lizzy. "Get Yer Yeah Yeah Yeahs Out." *The New York Times,* March 2013.

Graham, Paul. "Revenge of the Nerds," and "Hackers and Painters." *Hackers & Painters: Big Ideas from the Computer Age.* Sebastopol: O'Reilly Media Inc., 2004.

Guerra, Roberto, and Kathy Brew. *Design is One: Lella & Massimo Vignelli.* New York: First Run Features, 2012.

Harris, Mary Emma. *The Arts at Black Mountain College.* Cambridge: MIT Press, 2002.

Hazlitt, Henry. *Thinking as a Science.* New York: E. P. Dutton & Co., 1916.

Holmstrom, John, and Bridget Hurd, eds. *The Best of Punk Magazine.* New York: HarperCollins, 2012.

Hoover, Tim, and Jessica Karle Heltzel. *Kern and Burn: Conversations With Design Entrepreneurs.* Self Published, 2013.

Horn, Michael, and Clayton Christensen. "Beyond the Buzz, Where Are MOOCs Really Going?" *Wired,* February 2013.

Huizinga, Johan. *Homo Ludens: A Study of the Play Element in Culture.* Boston: Beacon Press, 1968.

Huxley, Aldous. *Brave New World.* New York: Harper & Brothers, 1932.

Isaacs, William. *Dialogue: The Art of Thinking Together.* New York: Doubleday (Random House), 1999.

Jackson, John. "Co-Teaching is More Work, Not Less." *The Chronicle of Higher Education,* February 2010.

Jonze, Spike. *Where the Wild Things Are.* Burbank: Warner Bros. Entertainment Inc., 2009.

JR. "My Wish: Use Art to Turn the World Inside Out." *TED,* March 2011.

Kaczynski, Ted. *The Unabomber Manifesto: Industrial Society and Its Future.* Livermore: WingSpan Press, 2009.

Kaku, Michio. *The Future of the Mind: The Scientific Quest to Understand, Enhance, and Empower the Mind.* New York: Doubleday (Random House), 2014.

Kelly, Kevin. "The New Socialism: Global Collectivist Society Is Coming Online." *Wired,* May 2009.

Keyes, Daniel. *Flowers for Algernon.* New York: Bantam, 1970.

Kirkham, Pat. "Lifelong Collaboration." *Charles and Ray Eames: Designers of the Twentieth Century.* Cambridge: MIT Press, 1998.

Klosterman, Chuck. *IV: A Decade of Curious People and Dangerous Ideas.* New York: Scribner, 2006.

Klosterman, Chuck. *Killing Yourself to Live: 85% of a True Story.* New York: Scribner, 2005.

Klosterman, Chuck. *Sex, Drugs, and Cocoa Puffs.* New York: Scribner, 2003.

Kreps, Daniel. "Girl Talk Unleashes Pay What You Want Album 'Feed the Animals.'" *RollingStone,* June 2008.

Kureishi, Hanif. "The Art of Distraction." *The New York Times,* February 2012.

Kushner, David. "The Flight of the Birdman: Flappy Bird Creator Dong Nguyen Speaks Out." *RollingStone,* March 2014.

Lasn, Kalle. *Culture Jam: How to Reverse America's Suicidal Binge—And Why We Must.* New York: HarperCollins, 1999.

Lasn, Kalle, and Adbusters, eds. *Meme Wars: The Creative Destruction of Neoclassical Economics.* New York: Seven Stories Press, 2012.

Lazar, Shira. "The Artists Behind The Gates Christo and Jeanne-Claude." *LX.TV,* July 2007.

Leadbeater, Charles. *We-Think: Mass Innovation, Not Mass Production.* London: Profile Books, 2008.

Lehrer, Jonah. "Groupthink: The Brainstorming Myth." *The New Yorker,* January 2012.

Lehrer, Jonah. "The Steve Jobs Approach To Teamwork." *Wired,* October 2011.

Leland, John. "Out on the Town, Always Online." *The New York Times,* November 2011.

Leonard, Elmore. "Writers on Writing; Easy on the Adverbs, Exclamation Points and Especially Hooptedoodle." *The New York Times,* July 2001.

Lessig, Lawrence, Ryan Merkley, Paul Brest, et al. *Creative Commons.* creativecommons.org, 2001.

Lewin, Tamar. "David Helfand's New Quest." *The New York Times,* January 2012.

Lewin, Tamar. "Instruction for Masses Knocks Down Campus Walls." *The New York Times,* March 2012.

Linn, Susan. *The Case for Make Believe: Saving Play in a Commercialized World.* New York: The New Press, 2008.

Lipman, Joanne. "Is Music the Key to Success?" *The New York Times,* October 2013.

Loy, R. Phillip. *Westerns and American Culture, 1930–1955.* Jefferson: McFarland & Company Inc., 2001.

MacFarlane, Seth. "Family Guy Viewer Mail #1." *Family Guy.* Season 3, Episode 21. Los Angeles: 20th Television, 2002.

Mann, Thomas. *Mario and the Magician.* Harmondsworth: Penguin Books, 1975.

Mendels, Pamela. "Can New Technologies Revitalize Old Teaching Methods?" *The New York Times,* March 1999.

Metz, Cade. "How to Build Your Own Google Docs (Without Google)." *Wired,* April 2013.

McAveeney, Corey. "An Unheard of Method of Collaboration: The Cross-Country CEO Swap." *Wired,* October 2013.

McCloud, Scott. *Understanding Comics: The Invisible Art.* New York: Harper Collins, 1993.

McCreight, Tim. *Design Language.* Portland: Brynmorgen Press Inc., 1996.

McLuhan, Marshall and Quentin Fiore. *The Medium is the Massage: An Inventory of Effects.* Corte Madera: Gingko Press, 2001.

McLuhan, Marshall, and Quentin Fiore. *War and Peace in the Global Village.* New York: Bantam, 1968.

McTeigue, James. *V for Vendetta.* Burbank: Warner Bros., 2006.

Miller, Claire, and Catherine Rampell. "Yahoo Orders Home Workers Back to the Office." *The New York Times,* February 2013.

Moore, Alan, and David Lloyd. *V for Vendetta.* New York: DC Comics, 1988.

Morris, William. *Political Writings of William Morris.* New York: International Publishers, 1973.

Newman, Judith. "If You're Happy and You Know it, Must I Know, Too?" *The New York Times,* October 2011.

Nyerges, Alexander. *Edward Weston: A Photographer's Love of Life.* Dayton: Dayton Art Institute, 2004.

Ojalvo, Holly Epstein. "Why Go to College at All?" *The New York Times,* February 2012.

Oswalt, Patton. *Zombie Spaceship Wasteland: A Book by Patton Oswalt.* New York: Scribner, 2011.

Palahniuk, Chuck. *Invisible Monsters.* New York: W. W. Norton & Company Inc., 1999.

Palahniuk, Chuck. *Lullaby.* New York: Doubleday (Random House), 2002.

Paperny, Vladimir. "An Interview with Denise Scott Brown and Robert Venturi." *Architectural Digest Russia,* 2005.

Pappano, Laura. "The Year of the MOOC." *The New York Times,* November 2012.

Pogrebin, Robin. "Couples Who Build More Than Relationships." *The New York Times,* April 2007.

Postman, Neil. *Technopoly: The Surrender of Culture to Technology.* New York: Alfred A. Knopf Inc., 1992.

Postman, Neil. *Amusing Ourselves to Death: Public Discourse in the Age of Show Business.* New York: Penguin Group, 1986.

ProjectProjects, eds. "Collaboration—All Together Now." *Print Magazine,* February 2011.

Rampell, Catherine. "A History of College Grade Inflation." *The New York Times,* July 2011.

Ravetz, Amanda, Helen Felcey, and Alice Kettle, eds. *Collaboration Through Craft.* New York: Bloomsbury Academic, 2013.

Rebello, Stephen. *Alfred Hitchcock and the Making of Psycho.* New York: St. Martin's Press, 1998.

Richtel, Matt. "Technology Changing How Students Learn, Teachers Say." *The New York Times,* November 2012.

Rivlin, Gary. "Leader of the Free World: How Linus Torvalds Became Benevolent Dictator of Planet Linux, The Biggest Collaborative Project in History." *Wired,* November 2011.

Robbins, Tom. *Jitterbug Perfume.* New York: Bantam, 1984.

Rock, Michael. "Designer As Author." *Eye No. 20,* Spring 1996.

Rock, Michael. "Fuck Content." *Multiple Signatures: On Designers, Authors, Readers and Users.* New York: Rizzoli International Publications Inc., 2013.

Rubin, Courtney. "Making a Splash on Campus: College Recreation Now Includes Pool Parties and River Rides." *The New York Times,* September 2014.

Rosenberg, Tina. "Beyond SATs, Finding Success in Numbers." *The New York Times,* February 2012.

Saarinen, Aline and Eero. *Aline and Eero Saarinen Papers, 1906–1977.* Washington DC: Archives of American Art, 1973.

Sadler, Simon. "TEDification versus Edification." *Places Journal,* January 2014.

Salinger, J. D. *Catcher in the Rye.* New York: Bantam, 1965.

Salonen, Essi. *A Designer's Guide to Collaboration.* designingcollaboration.com, 2012.

Sato, Julie. "Industry City, the Soho of Sunset Park." *The New York Times,* January 2014.

Scarfe, Gerald, and Roger Waters. *The Making of Pink Floyd: The Wall.* London: Orion Publishing Group Ltd., 2010.

Seinfeld, Jerry. *Comedians in Cars Getting Coffee.* Culver City: Sony Pictures Television, 2012–present.

Shaughnessy, Adrian. *Studio Culture: The Secret Life of a Graphic Design Studio.* London: Unit Editions, 2009.

Shea, Andrew. *Designing For Social Change: Strategies for Community-Based Graphic Design.* New York: Princeton Architectural Press, 2012.

Shirky, Clay. *Cognitive Surplus: How Technology Makes Consumers into Collaborators.* New York: Penguin Group, 2010.

Shirky, Clay. *Here Comes Everybody: The Power of Organizing Without Organizations.* New York: Penguin Group, 2008.

Shuster, Brian. "Virtual Reality and Learning: The Newest Landscape for Higher Education." *Wired,* December 2013.

Simmel, Georg. "The Metropolis and Mental Life." *Art in Theory 1900–2000.* Harrison, Charles, and Paul Wood, eds. Malden: Blackwell, 1992.

Solomon, Deborah. "O'Keeffe and Stieglitz: Intimacy at a Distance." *The New York Times,* August 2011.

Spiekermann, Erik. *Stop Stealing Sheep & Find Out How Type Works.* Third Edition. San Francisco: Adobe Press, 2014.

Stewart, Jon, Steve Bodow, Madeleine Smithberg, Lizz Winstead, and Chuck O'Neil. *The Daily Show with Jon Stewart.* New York: Comedy Central, 1999–present.

Stewart, Margaret. "Let's Stop Focusing on Shiny Gadgets and Start Using Tech to Empower People." *Wired,* September 2013.

Stimson, Blake, and Gregory Sholette, eds. *Collectivism After Modernism: The Art of Social Imagination After 1945.* Minneapolis: University of Minnesota Press, 2007.

Streitfeld, David. "Amazon Signs Up Authors, Writing Publishers Out of Deal." *The New York Times,* October 2011.

Tarantino, Quentin. *Pulp Fiction.* New York: Miramax, 1994.

Tate, Ryan. "Robot Professors Come With Singularity University's Massive Upgrade." *Wired,* August 2012.

Tolkien, J. R. R., and Christopher Tolkien, ed. *The Silmarillion.* New York: Del Rey Books (Random House), 1977.

Tsui, Bonnie. "Writing Alone, Together." *The New York Times,* July 2014.

Tzara, Tristan. "Dada Manifesto 1918." *Seven Dada Manifestos and Lampisteries.* London: Calder, 1977.

Thompson, Hunter S. *Fear and Loathing in Las Vegas: A Savage Journey to the Heart of the American Dream.* New York: Vintage Books, 1989.

Viction:ary. *So1o 2uo 3rio: Small Studios, Great Impact.* Hong Kong: Viction:ary Press, 2013.

Vonnegut, Kurt. *Slaughterhouse-Five.* New York: Dell Publishing (Random House), 1969.

Wallace, William. *Michelangelo: The Artist, the Man, and his Times.* New York: Cambridge University Press, 2010.

Walter, John. *How to Draw a Bunny.* Santa Monica: Artisan Entertainment, 2002.

Warren, Bennis, and Patricia Ward Biederman. *Organizing Genius: The Secrets of Creative Collaboration.* New York: Basic Books (Perseus Books Group), 1997.

Watterson, Bill. *The Calvin and Hobbes Tenth Anniversary Book.* Kansas City: Universal Press Syndicate, 1987.

Webber, Alan. "What's So New About the New Economy?" *Harvard Business Review,* January 1993.

Whedon, Joss. *Dollhouse.* Los Angeles: 20th Television, 2009–2010.

Wilde, Oscar. *The Picture of Dorian Gray.* New York: The Modern Library (Random House), 1992.

Wolf, Maryanne. *Proust and the Squid: The Story and Science of the Reading Brain.* New York: Harper Perennial, 2008.

Wolk, Douglas. "Future of Open Source: Collaborative Culture." *Wired,* June 2009.

Wood, Brian, and Becky Cloonan. *Channel Zero.* The Complete Collection. Milwaukie: Dark Horse Comics, 2012.

Zeldin, Theodore. *Conversation: How Talk Can Change Our Lives.* London: Harvill Press, 1998.

Zinn, Howard. *A People's History of the United States.* New York: Harper Collins, 2005.

RAS+E

Launching from the Graphic Design MFA program at the Maryland Institute College of Art (MICA) in Baltimore, Maryland, ras+e is a collaborative interdisciplinary design studio handling client and personal projects. They studied under Ellen Lupton and Jennifer Cole Phillips while contributing as writers and designers for *Graphic Design Thinking: Beyond Brainstorming.* Photographers, type designers, installation guerrillas, arts promoters, printmakers, culture critics, writers, activists, and professors; they teach design while slinging lo-fi messaging from Ohio, Florida, and Baltimore.

Some ongoing studio projects include: fashion graphics shop *re:print,* designing monospaces for physical applications through their foundry *re:type,* the Play series critiquing political action as entertainment, a collaborative manifesto on redesigning design education, and an activist comic exploration of tech and culture.

STATEMENT

Grafting and merging are boring. They happen in nature. Collisions are interesting. They require a particularly human capability for catastrophe. ras+e is a collision more than a partnership or a "my best + your best equals two brains better" sort of gig. ras+e works because the "+" is birthed, a third creature. Their collaboration is an entity unto itself—its own mode of making and thinking—its own agenda. They do not pursue Big Things through combined efforts; by slamming two designers together, they allow the being formed in the wreckage to become the creator. As famed thinker Franz Ferdinand once queried, "What's wrong with a little destruction?" Their mediums include Adobe Illustrator and Potassium Ferricyanide. Instead of justifying how these coexist, they break the one off in the other. And yes, then fire.